Health and Educational Scenario of Tea Plantation Workers

In the Context of The Plantation Labour Act, 1951

Dr. Mriganka Chakraborty

SPI PUBLICATIONS

Published by **SPI Publications**, www.spipublications.in

ISBN-13: 978-93-49860-82-7
ISBN-10: 93-49860-82-7

First Edition 2025

SPI Publications

Corporate & Editorial Office

117/2, Sector 37, Khandsa Road,
Gurgaon, Haryana 122 001
Phone: +91 98355 52940
Email: contact@spipublications.in
Website: www.spipublications.in

Processed and Printed in India

Preface

Tea industry plays a dominant role in the economy of India by providing employment opportunities to a hefty number of people in the country. Tea plantation was the first capitalist enterprise introduced by the British government in the middle of the nineteenth century. As the tea industry is labour intensive in nature, it needs a large number of labourers. At the very outset local labourers were recruited. As the local labourers were found inappropriate for performing tasks in the tea plantations, the management of tea gardens sought to importing the labourers from outside Assam. Since the British management of the tea gardens had prior experience of importation of labourers to their other colonies, they resorted to acquisition of labourers from the other parts of the country under the 'Colonial Policy'.

Since the beginning of plantations in Assam, condition of labourers was not at all satisfactory and this is revealed from the several consecutive strikes by the workers from 1840 to 1859. The situation was controlled only after the enactment of the Workmen's Breach of Contract Act, 1859. The introduction of indentured labourers in plantations further deteriorated the condition of labourers by reducing them from the status of free labour to a kind of serfdom.in tea plantations the indentured labour system was characteristically more akin to the feudal features rather than to the capitalist one.

The records of the Tea Labour Union of Cachar district of Assam reveal various facts relating to the treatment of tea labourers in the colonial period. In the later part of the nineteenth century the tea labourers were brought into Cachar district by giving false commitments. The coolies were told that they could earn money simply by shaking the bushes of the tea plants. Moreover, it has been pointed out that once the coolies entered the tea gardens, they were kept in such a way that they could not communicate with the outside world. Because the British management of the tea gardens thought that the coolies could run away from the tea gardens had they been allowed to contact with the outside world. The coolies while first coming to the plantations of the Cachar district, they were bound down by a contract to serve for a number of years. The coolies were subject to imprisonment for one month had there been breach of contract.

Since independence the Government of India has been taking various measures in order to improve the lot of these down trodden people. Plantation Labour Act, 1951 was passed in order to bring some changes in their life. There are many provisions in the Act for the improvement of the health condition, educational level, recreational facilities, etc. for the tea plantation workers. In spite of these measures, many plantation workers of a large number of tea gardens have not yet been improved at a satisfactory level.

Hence, a study to investigate the position of health, hygiene, educational level and welfare of tea plantation workers in the Barak Valley of Assam has been initiated along with examining the extent of implementation of provisions of the Plantation Labour Act, 1951, in the sample tea gardens. The study also focusses on the various measures of welfare as mentioned in the Plantation Labour Act, 1951 and their level of implementation.

The present study will be useful for the researchers who are working on the health and educational status of the tea labourers. The book is expected to provide valued research input to the researchers and academicians to undertake more exhaustive study on the problems relating to tea garden labourers.

Mriganka Chakraborty

Contents

Chapter I : Introduction

1.1. Background

Tea industry plays a dominant role in the economy of India. It provides employment to a hefty number of people in the country. It also constitutes as one of the major contributors to the foreign exchange reserve of the country. Tea plantation was the first capitalist enterprise introduced by the British government in the middle of the nineteenth century.

Robert Bruce and his brother Charles Alexander Bruce were the pioneers of developing tea plantations in Assam, India (Choudhury, 1989 and Bhuiyan, 1949). In 1839, Government started a tea garden at Chabua (near Dibrugarh, Assam, India) and another tea garden near Jaipur, Cachar, Assam, India on experimental basis with China tea plants and seeds.

In Barak Valley, first tea garden was set up at Barshangun (present Kathal Tea Estate, Cachar, Assam, India) in 1856 by G. Williamson (Verner, 1858) after the first discovery of tea plants in Cachar district in 1855. Thereafter, the number of tea gardens was gradually increasing year after year. In 1901, the number tea gardens in Barak Valley, Assam was 180 (Bhattacharjee, 1977). As per statistical data, in 2001 that number increased to 206 with 32, 272 hectares area under plantations. In 2005, the number of tea gardens remained the same (Tea Statistics, 2005). Again in 2007, the number of tea gardens in present Cachar district, excluding sick gardens was 104, which employed 31,200 permanent labourers and around 30,000 casual labourers. (Cachar Tea Labourers Association)

As the tea industry is labour intensive in nature, it needs a large number of labourers. At the very outset local labourers were recruited. As the local labourers were found unsuitable for performing tasks in the tea plantations, the management of tea gardens sought to importing the labourers from outside Assam. Since the British management of the tea gardens had prior experience of importation of labourers to their other colonies, they resorted to acquisition of labourers from the other parts of the country under the 'Colonial Policy' (Act VII of 1837). (Das, 2006).

In Assam, up to 1860, only the local labourers were recruited in plantations. In 1859 the number of local labourers was 10,000 although the labour requirement was about 16,000 to 20,000 hands. After 1865-66 due to the rapid expansion of plantations, labourers were arranged from Bengal, Bihar, Orissa, Madhya Pradesh, Andhra Pradesh and present Chennai. The number of imported labourers was 40,000 in 1872 and 226, 321 in 1880. In 1921, about half a million tea labourers were from outside tea plantations. In 1931 there were about 1.4 million tea labourers of which 9,00,000 were in tea plantations and 5,00,000 were in the neighbouring villages for living on agriculture. After 1931, the import of tea labourers gradually declined and was totally stopped by 1941. By 1941, the total number of tea labourers rose to 10,00,000 (Annual Report on labour immigration into Assam, 1879-1905). In 1971 there were 22,40,000 immigrants in Assam (Report on Immigrant Labour in The Province of Assam, 1932-33).

As per the statistics, among the imported tea labourers in Cachar, Hindi Communities constitute 5%. Majority of the tea labourers in the gardens of Cachar originated mainly from the states of West Bengal, Bihar, Jharkhand, Orissa and Andhra Prodesh. Near about 20 caste or sub caste people, namely, Bhor, Rajbhor, Bhumiz, Nunia, Mal, Robidas, Dhubi, Roy, Dusad, Basfor, Tanti, Rikiason, Ree, Bauree, etc. migrated from the districts of Purulia, Bankura, Bardhamaan, Medinipur and Birbhum of West Bengal, India. In the same manner, the castes like Tantubai, Nayak, Chasha ,etc. originated from the state of Orissa, India. Besides, some tea labourers originated from Jharkhand and Andhra Pradesh. Present tea labourers who are working in the tea gardens are the descendants of the coolies migrated in the colonial period. Thus the people of diverse socio-cultural background continued to stay together in the plantations of Cachar district as wage labourers over the period of time.

1.2. The system of Recruitment of Tea Gardens Labourers in the Colonial Period

At the beginning of the tea industry in Assam, local labourers were recruited, but after a few years indentured labourers were employed through contractors. The Government passed the Transport of Native labourers Act of 1863 for licensing recruiters and to regularise the recruitment system. The contractors were known as Arkattis. The Arkattis were the recruiting agents of the planters and they were in an advantageous position to identify sources from where the labourers could be recruited. The poverty ridden Tribal people were the main target for the Arkattis because those people could not find any other means of livelihood in any way. (Guha, 1977).

The labourers were transported at the expense of the planters. The Act was amended in 1870 when the Sardari system of recruitment was recognised. However, the earliest system of recruitment through the Arkattis was not stopped. The planters practised the two systems of recruitment till 1915. Under Assam Labour and Immigration Act of 1955, recruitment through Arkattis was made unlawful and Sardari system remained as the only system of recruitment in tea plantations. In 1917, the Tea District Labour Supply Association was formed to control and coordinate recruitment under the Sardari system. Since then this was the only organisation concerned with the recruitment supervision. The problems of unemployment of the immigrant labourers became the important factor for not favouring further migration of labourers into Assam. After 1941 the importation of labourers from outside Assam was stopped but again in 1955 about 11,000 labourers were recruited from outside. From 1960 onward, recruitment from outside Assam was totally stopped and the Tea District Labour Association was abolished. Then, labourers were employed from the existing unemployed labourers within the state. (Allen, et.al.,2001)

1.3. The Condition of Plantation Workers during Pre-Independence and Post-Independence Period

Since the beginning of plantations in Assam, condition of labourers was not at all satisfactory and this is revealed from the several consecutive strikes by the workers from 1840 to 1859. The situation was controlled only after the enactment of the Workmen's Breach of Contract Act, 1859. The introduction of indentured labourers in plantations further deteriorated the condition of labourers by reducing them from the status of free labour to a kind of serfdom.in tea plantations the indentured labour system was characteristically more akin to the feudal features rather than to the capitalist one. Up to 1901, the indentured labourers used to get wages of Rs 3 per month, while the minimum wages was set by the Government were Rs 5, Rs 4 and Rs 3 for men, women and child workers respectively. But the free labourers of the organised sectors like P.W.D. and Railway used to get higher wages, i.e. Rs 7 per month. (Guha, 1977 and Report on Tea Culture in Assam).

Up to 1908, the planters were empowered to arrest, without warrant, those workers who escaped from their plantations and the planters enjoyed this privilege up to 1908. This private power of planters to arrest the escaped labourers was abolished in 1908. The Act of 1901 already mentioned that a time expired labourer could not be compelled to accept another contract. However, this could not stop alluring of time expired labourers and they were forced to accept another contract after the expiry of the earlier contract by the planters. But the penal labour contract under the Workmen's Breach of Contract Act, 1859 was being practised. But it was modified only in 1920.

The way the labourers were recruited and brought into the plantations was brutal. In 1861, approximately 89 percent could reach the plantations. The remaining either died on the way due to inhuman treatment or affected sickness. Labourers including men, women and children were enticed and treated like slaves and they were arrested whenever they ran away. Due to the unhygienic condition and inadequate food many such labourers expired on the way to plantations. As the labourers were forced to accept another contract after the expiry of one contract illegitimately, they could not return to their native villages. They were not provided with proper houses, sanitation facilities, drinking water, medical facilities, etc. due to fear of imprisonment, the labourers could not call strike or stop the work. Whipping which prevailed in feudalism to control the slaves was the common practice in these plantations. Also, the labourers belonging to one plantation were not allowed to meet their friends, relatives in other plantations without the permission of the plantation managers. In plantations, labourers were more or less treated like serfs, attached to the same as they were by extra economic means. To have assured supply of labourers, the labourers were tied to plantations through contract and by offering land. They were sometimes allotted small plots of land for paddy cultivation. Due to poor economic condition of labourers, they were sometimes compelled to borrow money from the manager. This again created further domination as this loan was to be repaid from their wages. These were the feudal features of plantations of Assam, which closely resembled those of the 'latisfundismo' agricultural estates of 'Latin America'. (Feder, 1971).

The records of the Tea Labour Union of Cachar district of Assam reveal various facts relating to the treatment of tea labourers in the colonial period. In the later part of the nineteenth century the tea labourers were brought into Cachar district by giving false commitments. The coolies were told that they could earn money simply by shaking the bushes of the tea plants. Also it has been pointed out that once the coolies entered the tea gardens they were kept in such a way that they could not communicate with the outside world. Because the British management of the tea gardens thought that the coolies could run away from the tea gardens had they been allowed to contact with the outside world. The coolies while first coming to the plantations of the Cachar district, they were bound down by a contract to serve for a number of years. The coolies were subject to imprisonment for one month had there been breach of contract.

The coolies were generally provided with basic necessities of life like rice, wheat, kerosene, etc. by the management of tea gardens in the colonial period. Also, the coolies were given the necessary materials for the construction of their houses. In the later part of nineteenth century the tea labourers had to work against a wage of 4 paise to 6 paise per day. Again due to illiteracy the coolies could not live a hygienic life and as a result they had to suffer from various deadly diseases like diarrhoea, malaria, tuberculosis, etc. During colonial period, in the tea plantations, there were no provisions for the treatment of the coolies. As a consequence, they were the worst sufferers.

Although, later on the British management set up hospitals in the tea gardens but in most of the cases due to lack of awareness of the coolies, they were mainly dependent on the superstitious beliefs for coming around such deadly diseases (Tea Labour Union Records, Cachar, Assam). The British management considered the coolies an instrument of producing money. They also encouraged the coolies to have wine in the evening or after the working hours. The management of the tea plantations always wanted that the labourers should not be allowed to think of their further development (Chottopadhay, 1989).

The earlier paragraphs reveal how the tea labourers were brought into the thraldom in the colonial period. But till today the tea labourers are not fully freed from the clutches of exploitation. Also, the labourers in the Barak Valley are provided less wages per day than the labourers in the Brahmaputra Valley, Assam. in some of the tea gardens of Cachar district the workers are forced to work at a rate less than the specified rate. Sometimes rate of wages paid to the permanent workers are higher than the casual workers. Though the discrimination is made with regard to payment of wages, yet the working hours are same for both the types of workers which is eight hours. Also, the casual labourers can not avail free medical treatment from the garden hospital.

1.4. Review of Literature

A number of Studies have been conducted by various authors on the diversified aspects of tea gardens and labourers of tea gardens. Socio-economic conditions of tea plantation labourers also received a special attention. In the following paragraphs a brief review of such studies has been provided.

Awasthi, (1975) identified various additional facilities and fringe benefits provided by the management of tea gardens towards the workers. These included free housing, free medical facilities, free education to children up to primary age, establishment of recreation centres, providing free tea to the workers on the work, sickness allowance, leave with wages, etc

Harlalka (1975) identified the socio-economic problems of the tea garden labourers of Assam. Bhattacharjee (1977), while elucidating the history of tea industries in Barak Valley, made a brief review of the financing patterns of tea industries in the colonial period where it was specifically mentioned that the British traders were the pioneers in providing the financial aid to the owners of the tea gardens. An introduction about tea, the history of it, the main tea growing countries, etc. are given by Ashby (1977). Ashby gives a detailed description about the drying and processing of beans and leaves and about the by-products which can be developed in the process. The details of planting, fostering, manufacturing, and pest control measures in tea cultivation are also explained in the book. Kar, (1978) explained marriage systems of one of the migrated groups working in the tea plantation of Assam. Data in respect of marriage was collected from all the 164 couples present in the garden at the time of survey. It was found that the highest incidence of married male and female were in the age group of 30-35 years and 25-30 years respectively. Goradia (1979) made a study on strength, weakness, opportunities and threat (SWOT) analysis of Indian tea industry. According to him the strengths of tea lies in its caffeine content which makes it a stimulating drink; it is economical; and it is a self reliant agro industry; and India is the largest consumer of tea. The weaknesses are dependence on vagaries of weather, high cost of production, fluctuations in quality, labour intensiveness and long gestation period. Opportunities are the home market which provides an excellent base for product sophistication such as instant, cold and fruit mixed teas. Another opportunity is that tea and coffee are interchangeable. The threats are prolonged slumps in prices and undue rise in cost of production. The study also deals with different aspects of tea such as tea tasting, auction, tea research and marketing. Goradia gives a detailed picture about plantation technology for growth. According to him plantation technology is a synthesis of industry and agriculture. Economies of scale are prevalent in tea industry and this factor hinders the small fanners from having research on production, manufacturing and marketing, and from training of workers. Dutta (1983) discussed the problems and prospects of literacy and employment situation of the tea labourers of Assam. Thakurta (1986) highlighted the role of women workers in India's tea industry. The paper hinted the rate of wages paid to male and female plantation workers in the states of Assam and West Bengal.

Socio-economic conditions of the South Indian tea plantation workers were examined by Raman (1986). Historically, colonization paved the way for the formation of plantation system in India During the early period, standard of living of workers in plantations was very poor. They were ill-fed, ill-housed and ill-treated, the researcher observes. Since independence, the constant protest of these labourers resulted in the introduction of Plantation Labour Act 1951. Raman also analysed the labour market In Kerala and Tamil Nadu there are two types of workers, permanent and temporary. But in Karnataka, in addition to these permanent and temporary workers, casual/contract workers are also engaged. An important factor noted by the investigator is that productivity of women workers is high. Living conditions of workers are poor with inadequate drinking water, poor housing facility and insufficient medical care.

Tea plantation industry is a labour intensive one and majority of workers are women. Concentrating on the role of women workers in trade unions Sarkar and Bhownik (1988) made a study on West Bengal tea plantation industry. They found that participation of women in trade union activity is low, and the major reasons, according to them, are inequality, low literacy rate, low political consciousness and burden of the household duties. Bhowmik (1988) highlighted the importance of the institution of co-operatives in Indian tea industry by making a study on Allied Plantation Worker's Co-operative Limited in West Bengal and Tachai Workers Co-operative in Tripura. He observed that worker co-operatives are based on democratic principles and it encourages and make provision for workers participation in decision making. This helps to prevent concentration of power in bureaucracy. Apart from sharing of benefits among members, cooperatives are working with the larger objective of welfare of the society as a whole, the researcher observes.

Chottopadhay, (1989) highlighted various brutal exploitations over the tea labourers in the colonial period. The study highlighted that the tea labourers were deprived from the welfare facilities and the British management could not think of having any provisions of safety and welfare measures for the tea workers. Kurian (1990) studied the socio-economic background and consumption pattern of women workers in tea plantation industries in Munnar, ldukki District in Kerala, The study was focussed on the personal and family background of women workers, their economic and living conditions, income and expenditure pattern, standard of living, social status, working conditions, welfare facilities and security. The tea industry is labour intensive and majority of workers are women.

Baak (1992) studied the historical evolution of plantations in Travancore. In Travancore, plantations were developed later in 1860s compared to Ceylon, Assam and Bengal where the process started in 1820s, 1830s and 1840s respectively. He points out those plantations are developed with political influence from the Colonial Government, and the pressure came to Travancore was from the British through Madras Government during 1860s. Initially, the main investment was for construction of roads. The British owned the plantations and the labour came from the backward communities. With the ownership rested with the British, profit during the colonial period went out of the country. The positive aspects of development of plantations are development of infrastructure such as roads, transport and communication facilities, increase in economic activity and increase in employment opportunity. Goswami (1992) explained the socio-economic life of the ex-tea garden labourers. The study analyses the socio-economic status by stratifying the persons engaged in different occupations viz. agriculture, partly in agriculture, others and miscellaneous. The work is basically a study according to the strata of occupations and is mainly confined to the Brahmaputra valley of Assam.

Jain (1995) also points out the positive aspects of tea drinking. He argues that tea drinking helps to prevent cancer based on the observations made in the Annual conference of the American Association for Cancer Research (MCR). Subhramaniam (1995) presented various information like genetic potentials of the bush, training of plants and management of cultural inputs to achieve optimum results, maintenance of the bush etc of this plantation crop in a lucid manner. Besides the relationship between the productions, domestic consumption and exports of tea has been depicted. The statistical data relating to the trend of tea exports, different exports markets and the revenue earned up to the year 1993 has also been provided.

Bhadra (1997) in his study tried to highlight some features of tea plantations and surplus labourers in Assam. He presented a detailed picture of the growth of tea plantations in Assam. The paper narrates the story about the migration of indentured labourers into the plantations i.e., how the labourers were recruited and their conditions. Moreover from his study a brief picture about the organizational structure of tea plantation in Assam and West Bengal can be assessed. He tried to represent the socio-economic condition of the surplus plantation labourers in Upper Assam. As against the general observations of under production,

Chakraborty (1997) warns against the possibility of oversupply of tea in the world market and a depression in international price. The disintegration of the USSR and United Nation's economic blockade on Iraq affected Indian exports adversely. In export, India's position slipped to fourth position, but in value terms India fetches maximum price due to superior quality. In order to overcome the threat of oversupply internal consumption may be increased further. Around 95 per cent of the Indians consume tea. Consumption of tea is inelastic to income. Chakraborty (1997) presented a detailed picture relating to the socio-economic development of plantation workers in North East India. It is revealed that the economic condition of the workers in Tripura was apparently worse than the workers in Assam as the tea gardens flourished under the patronage of private individual and companies. Moreover, in this study an attempt has been made to demonstrate in details the role of education in the development of human resources, the concept of workers education, and the role of International Labour Organization (ILO) to promote workers' education all over the world, the workers' educational programmes in India.

Kar (1997) described the health and sanitation among the tea labourers in Assam. He also examined the level of implementation of the provisions relating to health and sanitation as mentioned in the Plantation Labour Act, 1951, in some of the tea gardens of Assam. He found that in most of the tea gardens in Assam these provisions were not properly implemented and the health facilities were far from satisfactory level. It was revealed from his study that plantation labourers are found to suffer from a large number of diseases like- gastro-enteritis, respiratory trouble, anaema, dysentery, etc.

In India research and development activities are undertaken for developing technologies for tea plantations. Muraleedharan (1998) gives a description about the innovations in this aspect by UPASI. They have developed two types of pruning machines and two models of plucking machines. But the machines have some drawbacks such as high weight, overheating, noise and problems related to technical services and spare parts. Dwibedi (1999) conducted a study on tea plantation industry in West Bengal. Growth trends in tea production, area and yield and influence of climatic factors like rainfall, temperature and humidity are also taken into account. The study revealed that area under tea in West Bengal has been increasing more or less at a constant percentage rate between 1961 and 1993. Production of tea has been increased with a declining rate of growth and yield rate increased with a declining trend. The analysis revealed that rainfall and temperature effects are dominant in crop production in West Bengal, particularly Dooars. His examination of the size of plantation and type of ownership and *its* impact on productivity showed that there exists a positive relationship between size and yield; and the propriety ownership is found to be less efficient The analysis on acreage and yield response to price suggests that planter's decision in expanding acreage under tea in any year is influenced by the magnitude of the difference between the price realised and the price which the planter expects to realise. He concluded that decision to bring more area under tea is dependent on the last two year's price level.

Debnath (2000) dealt with the dynamics of social realities, the continuity of cultural products, the interactions between the two and the emerging social patterns of life in a settlement village originally populated by the transported labourers and presently by their descendants in the tea garden. For collection of data, a structural schedule was administered to 100 households and informal interviews and observations were extensively used. Laskar (2001) emphasized the problems regarding the entry of inferior tea into our country from neighboring countries and observed that Indian premium quality Tea is getting damaged in terms of its image. He suggested that for boosting the market of tea nationally and internationally and health effects of tea need to be thoroughly published.

Sri Lanka Tea Cluster Supported and funded by The Competitiveness Initiative, a joint project of the United States Agency for International Development (USAID), Nathan Associates Inc., and J.E. Austin Associates (2002) made a study on the competitiveness of Sri Lanka's tea industry. In the study attempts had been initiated to develop a competitiveness strategy for Sri Lanka's tea industry. To become more competitive, the industry must build on its historic strengths to develop brands, form marketing alliances, and stir closer to the end-consumer in its target markets. Doing so will help guard the industry from price volatility while creating value in the market. In response to these issues, members of the tea industry have devised a strategy for repositioning the industry in world markets. This strategy will be pursued through five strategic initiatives, viz.,

- Position the Industry Closer to the Consumer
- Forge Alliances with National Marketers of Ceylon Tea-rich Blends
- Launch a Rediscovery of Ceylon Teas
- Develop the Workforce
- Restructure the Public–Private Partnership.

Goswami (2002) stated that tea industry was the main economic force behind the urbanization in upper Assam and in the lower Assam due to slow growth of tea-industry, the process of urbanization was low. He found that the tea industry in Assam was developed by a handful of

British entrepreneurs with Government patronage and support with the objective of maximizing the returns on capital employed. The study indicated that the shabbily treatment and exploitations of labourers in the initial period. Finally the paper observed the positive effects of tea gardens in the economy of Assam. Shil (2002) undertook a study to find out the marketing practices followed by the Indian tea industry with particular reference to the industry's handling of the marketing mix and the measures taken by the industry to re-orient its marketing practices during the period 1991-2000.

Wickramasinghe and Cameron (2003) found that socio-cultural and political relations dominate tea production and as such the theory of economics of scale has a loose interpretation in the context of Sri Lankan tea industry. The continuity of the plantation holdings depends on raising productivity of land and labour. The major challenges for policy makers are to raise the profitability through raising productivity and consequently lowering the cost of production. However, this sector is challenged by problems of shortage of labour and land, adequacy of capital, inconsistent product quality, exploitation by intermediates, and marketability of tea leaves. Ariyawardana (2003) in an intensive study on value added tea producers in Sri Lanka examines sources of competitive advantage and studied how it relates to the performance of the tea growers. His study provides a deep understanding of this issue from the management point of view. Sharma (2003) studied the industrial relations, socio-economic life of the Temi tea plantation workers in Himalayan region. The study suggested that industrial cohesion between labour and management in only ultimate alternative for the survival of Temi Tea Estate in the near future.

Sharma (2005) addressed the issues like social transformation, occupational mobility, cultural integration, inter-ethnic relationship among the tea garden labour communities of Anula Shagun Tea Estate of Arunachal Pradesh. Thapa (2005) studied the tea industry of Nepal and presented various aspects like prospects of tea industry in Nepal, like- strategy for the promotion of Nepalese tea, position of Nepal tea industry, etc. He made a brief review of the financing mechanism of tea industry in Nepal and identified that the major source of loan for development of tea plantation and for the establishment of processing units for both large estates and small farmers is the Agriculture Development Bank (ADB/N). Though tea sector has a promising future, its financial performance has been weak and most borrowers have not been able to pay interest and principals on their borrowed money. The paper mostly concentrated on conceptual matters rather than identifying the problem of tea industries.

Singh, Narain and Kumar (2006) provided a detailed picture of historical background, recruitment, socio-economic profile of tea workers including political participation of tea workers in Cachar district of Assam. Choudhury (2006) highlighted that Assam tea is earning around 50 per cent of the foreign exchange earned by India's tea industry, although its demand is already in recession due to better quality-tea supplied by countries like Sri Lanka, Cuba, etc at comparatively lower prices. It was also highlighted that the most serious ailment remains not only in low productivity but also with quality of produce due to low investment on infrastructure and low managerial efficiency and it was suggested that high cost of production and stagnant productivity needs to be addressed on an urgent basis to overcome the problems. Das (2006) brought into limelight a brief historical picture relating to deteriorating conditions of tea industries in Barak Valley of Assam. He mentioned that employers are denying various welfare measures to the workers on the ground of reduction of international prices of tea. Kumar (2006) in a study of immigrant tea plantation labourers has covered literacy, income, employment and expenditure indicators of socio-economic life of ex-tea garden labourers. Saikia (2006) in his article discussed about the prospects of tea tourism in Tinsukia and Dibrugarh districts and he also mentioned that tea tourism could be explored by tea planters as an alternative source of revenue, for the tea industry hard hit by falling international prices, stiff global competition and labour problems. He has also mentioned about the various features of tea tourism.

McLeod Russel India Ltd. & Harrison Malayalam (2007) made a study on the various aspects of Indian Tea Industry viz., importance to economy, export growth, growth potential of Indian Tea in Russian Market, etc. The study also reflected that Indian tea industry is highly sensitive to the Government Policies and is highly dominated by organized players.

Huque (2007) tried to devise the value system of the tea industries of the developing countries and associated problems and side by side attempted to underscore the strategic solution in the light of the strategic accounting model of the developing countries. The study further emphasized on the issues which affect the developing countries tea industry and an alternative solution to the problem. Two countries, viz, Bangladesh and Japan were selected for the purpose of the study. The study found that small tea growers in Japan get funds from cooperative association which is not prevalent in Bangladesh but sanctioning and disbursement of loans by banks to tea planters in Bangladesh mostly depends on the recommendations of Bangladesh Tea Board.

Chakraborty (2008) made a study on the health, education and welfare facilities of the tea workers of Cachar district of Assam in the context of the Plantation Labour Act, 1951. The study mainly focussed on the level of implementation of the various provisions relating to health and welfare of the said Act in the tea gardens of Cachar district. Findings of the study revealed that the management of the tea gardens under study is not properly implementing the provisions of the Act. Wal Van Der (2008) studied sustainability issues in tea sector and made a comparative analysis of six leading tea producing countries of the world. The study found that working conditions for pickers are often poor, with low wages, low job and income security, discrimination along ethnic and gender lines, lack of protective gear and inadequate basic facilities such as housing and sometimes even drinking water and food. At the same time there is no possibility for tea plantation workers to improve working conditions because trade unions are ineffective or absent and/or are not representing them because most of them are temporary workers. While tea production by smallholders is growing worldwide, their situation is often problematic because the prices they are paid for fresh tea leaves tend to be below the cost of production, among other factors.

Banerjee and Banerjee (2008) in their book "Global Tea Trade: Dimensions and Dynamics" systematically described the global tea trade scenario. The book not only concentrated on the mechanisms involved in the production, marketing and exports of tea for major tea producing countries of the world but also highlighted import pattern and re-exports of tea in major non-tea producing countries. In this process, efforts had been made in the book to suggest policy issues based on field data collected, compiled and analyzed. Dasgupta (2008) analyzed the impact of New Economic Policy on the tea industry of Assam. The analysis was divided into two parts. Part I dealt with analysis of some parameters such as annual tea production, yield rate per hectare, area under tea plantations, number of tea estates, income, employment and also sickness to understand the impact of new economic policy on the tea industry of Assam. The part II mainly reflected the conclusions and suggestions. In a similar study, Das (2008) discussed the impact of globalization on the tea industry and silk industry of Assam. She took the parameters like price per kg, quantity of exports of tea to expose the impact of globalization on tea industry of Assam. She suggested that the tea industry of Assam must accept the strategies like export promotion, diversification of marketing portfolio, value addition to the product, cost reduction, etc. to meet the challenges posed by globalization.

Athparia (2009) studied the settlement and socio-economic life of the tea garden labourers in Cachar district of Assam. He investigated and found that the present tea garden labourers are the fourth generations in the tea plantations. The functional role of the family of the tea labourers have changed. He also found that the conditions of living in the tea gardens compelled the labourers to give up their traditional norms and customs. Baruah (2009) examined the health, hygiene and sanitation among the tea garden labourers of Jamira Tea Garden, Assam. It was found that tea labourers were mostly prone to infectious and water borne diseases due to poor sanitation and hygiene. Balgir (2009) provided a brief overview of the health and morbidity pattern in the tea garden communities of North East India. The study found that genetic disorders are gaining prominence and have profound health implications in the morbidity status of tea garden labourers in Assam. Nagoor (2009) examined the performance of India's tea exports and figured out that tea, which was a major exportable commodity in India's agricultural exports, is declining steadily. The export performance over the last three decades shows that the percentage share of India's tea export in total world tea exports has declined drastically. During 1981-90, the share was 21.91%, which declined to 13.35% in 2001-2004. The decade of the nineties and beginning of the twenty first century has been quite depressing for the tea industry in India. Sengupta (2009) studied on the measures taken by the Tea Board of India to make tea the popular choice over soft drinks and all other beverages. Ghosal (2009) in his article tried to reflect various issues relating to tea viz, affect of monsoon on tea production in the first half of the year 2009, push up in the global prices of tea due to shortage, areas of concern for tea, effect of recession on tea export, Tea Boards plan to overcome the problems of tea export, etc.

Hussain and Hazarika (2010) analysed the crisis of tea industry of Assam. According to their study, a considerable number of teagardens of the State had gone sick over the period due to lack of infrastructure, modernization and efficient management. Moreover, India's tea market was facing yet another paradox which could be explained in terms of glaring gulf between the price received by producer and the price charged by dealers and retailers. The common consumer in the market was confused of the fact that while the producers were facing the crisis created by a market accumulation and decline of prices, often voiced by the corporate, the benefit of low price did not come to the common consumers. The reason perhaps slandered in non-conformity with regulated market behaviour of producers among whom many were found to be selling out their produce directly without routing it through auction centres. They suggested that the tea industry should get modernized with a change in technique of plantation, improvement of infrastructure, encouragement to the electronic tea auction and managerial excellence. It was also recommended that a separate small grower Development wing within Tea Board should be set up. Mitra (2010) analysed impact of globalisation on industrial relations on tea plantations of Dooars region of West Bengal. The study emphasised on establishing a harmonious relation in the tea industry for its overall growth and development.

Haviland (2011) in his article highlighted the rising costs, falling prices and poor weather were making the production of tea in Sri Lanka more expensive. He also suggested that the tea industry in Sri Lanka should adopt the policy of product diversification to survive in this condition.

Hazarika (2011) in her article highlighted the changing market scenario for Indian Tea. In international scenario, Indian tea is bringing up the rear due to high price and poor quality. In domestic market too tea is facing tough competition from soft drinks and other beverages. She mentioned that traditional tea industry was production oriented but in the present scenario to meet the competition, a new market orientation is required. The market production linkage could be strengthened by either Market creation or Market matching. Market creation refers to an attempt to create new markets. In market matching there was very little effort to influence the market. A product was sold to match the market demand. Few years back tea industry had the market matching approach. But according to her a combination of both market creation and market matching is needed to face the demand of the market. Harizan (2011) took a historical glance at the emergence of tea industry in Darjeeling district and analysed its current problems with a view to retaining the erstwhile glory of Darjeeling tea. Being a labour-intensive industry, the author proposed that the potential for development of human resources of tea industries was rich, especially by enabling the upcoming generation of tea garden population in moving on and even in sharing in the ownership of the garden.

A study on the Total Factor Productivity Growth (TFPG) of 15 selected tea gardens of upper Assam for the period 2000 to 2009 has been made by Deb, Roy and Mazumder (2012) on the basis of panel data on plantation sector inputs and output. TFPG is measured using the Solow Divisia Index. A Cobb Douglas Production Function is estimated in order to find the output elasticities with respect to inputs. From the study, it is found that area under cultivation and implements are the only two significant factors that explain output. From bivariate correlation analysis it is clear that pesticides and fertilizers, mature cultivation and labour are pair wise highly positively correlated. The study concludes that inputs have grown at a very high rate, outpacing the rate of output growth. Almost all the gardens exhibit negative TFPG each year with few exceptions.

In a study entitled 'A modified approach to bringing up of young tea into bearing' Goswami and Dutta (2012) highlighted that sustainable high productivity in tea depends primarily on type of bush frame formed during early years of its establishment. This can only be obtained by pruning of the branches at the right time. The experimental results suggested that two frame forming prunes on the softer wood can help in obtaining a bush frame with high yield potential. However, to do so high tipping not below 75 cm must have to be adopted from decentring onward. The performance of young tea may vary depending upon planting material, soil health and other growing conditions.

Chakraborty (2012) made a study on the profitability of tea gardens of Barak Valley of Assam. It was revealed from the study that tea gardens in that region were sustaining losses due to increase in cost, decrease in average auction prices of Barak Valley tea and mounting burden of tax (cess & excise duty) on tea. Kar and Mathew (2012) made a micro analysis of tea industry in West Garo Hills District of Meghalaya. The study had been undertaken to examine the evolution and present status of tea industry in the district of Garo Hills. They found that tea industry in Garo Hills is exhibiting relatively a high productivity (5.4 ton per hectare) and has an immense potential as an economic crop. The study also highlights the intricacies and challenges faced by tea industry in the region.

Sundar (2012) narrated in detail about the initiatives to encourage quality production in the tea industry in the South India. He also mentioned that these initiatives launched in September would have definite bearing on India's overall tea trade and were fit for emulation by the tea sectors in different parts of the country.

Arya (2013) identified that major factors responsible for poor performance of Indian tea Industry were towering cost of production, the old age of tea bushes, lack of infrastructure, high price, labour problem, inefficiency of the Tea Board, rising cost of labour, etc. Incapability of the Indian tea to face competition in the international market, it is bit by bit bringing up the rear its time-honoured buyers and the liberalization phase has been less flourishing in revitalizing the tea growers. Sharma (2013) analysed the socio economic condition of tea garden labourers of Lohpohia Tea Estate of Jorhat district, Assam, India. He found that general living conditions of the labourers are poor and unhygienic and housing condition of the tea labourers is very pathetic. The labourers also use open space for defecation and other call of nature in the present globalised era.

Devi (2014) examined the socio-economic status of the tea garden women workers in Assam, India, with special reference to sonitpur district. The major findings revealed that there were various drawbacks in the Tea garden women workers.

Bose (2015) highlighted a brief history of tea plantation workers in Assam and found that the working condition of the tea plantation is still poor. Chakraborty and Barman (2015) studied the potentialities of India tea trade with East-Asian countries in the context of the Look East Policy. It has been found that the proportion tea trade of India has been lower than the rest of the world. Chakraborty and Ahmed (2016) examined the institutional financing of tea industry in India. The study found that the proportion of bank finance towards the tea industry has been lower than that of other industry. But the correlation between bank finance and production of tea industry has been significant.

Devi (2022) examined the socio-economic conditions of tea garden workers in Biswanath district of Assam. It was revealed from her study that most of the tea garden workers were brought from the mid 19th century to the mid 20th century from the tribal heartlands of Central-Eastern India as indentured labourers. She found that conditions of the most of the tea garden workers of Assam are very disappointing until today. She explored the socio-economic conditions of tea garden workers in the various tea estates of the Biswanath district of Assam.

Hazarika, Ozah, Satyakam, Pegu and Gogoi (2024) tried to evaluate the socio-economic conditions of tea garden labourers in the Assam. Data were drawn from 400 respondents to understand different factors determining the wage, which includes district differences, gender, educational status, work experience, and household size. The study found significant differences in the mean wages, household income, and consumption across the districts. It also highlighted the gender wage differentials where males earned more than females, and higher wage levels were influenced positively by higher education levels, greater work experiences, and larger household sizes. The findings revealed the implication of interventions targeted at the economic enablement of tea garden workers, especially regarding gender inequality and opportunities in education.

From the aforesaid review of literature, it is clear that tea plantation industry has an important place in economic activities all over the world. Moreover, Tea is considered as a global stimulating drink. Most of the tea plantations in the world have been developed during the colonial period. Hence, an element of exploitation has historically been associated with tea. India is not an exception to this. In India, tea blooming areas are North East India and south India. Many of these tea plantations were developed during the British period. Since the independence, interest of the British shifted towards the African countries and subsequently the Indian tea plantations began to undergo. Moreover, It is revealed from the aforesaid studies that most of the studies relate to importance of tea industry, tea garden labourers, tea production, tea marketing, etc. In addition, it is revealed that there have been rare studies based exclusively on examining the health, hygiene, education and welfare scenario of tea plantation workers along with the level of implementation of the provisions of Plantation Labour Act, 1951 in the Barak Valley region of Assam.

1.5.Statement of the Problem

Thus, during colonial period the workers had to live under ruthless conditions. They had to work for long period of time without rest and were brutally being exploited by the management. A significant part of the tea labourers was illiterate. They suffered from various communicable and waterborne diseases. The death rate, infant mortality rates were much higher than that of the general population. The working class in the tea plantations of Assam is perhaps the most exploited in the organised sector of the economy. Low wages, poor housing and lack of avenues for social mobility have been a recurring theme since the inception of the plantation in North East India in the early 19th century. Among the total tea garden working labour in each tea garden, only 30–40 per cent of them are permanent employees. During the peak season, each garden employs casual workers at wages much lower than the actual minimum wage. There are no maternity benefit schemes for the tea garden workers. It has been observed that during pregnancy and post natal period, women continue to engage in hard jobs. Most allegations of child labour in the tea industry involve the functions of plucking, weeding, hoeing, and nursery work. And poor socio-economic conditions, ignorance due to illiteracy, over-crowded and unhygienic living conditions in the residential colonies make tea garden population helpless to various communicable diseases and underfeeding. The issues relating to open defecation and lack of drainage systems need to be addressed at the earliest. Regulatory provisions states that tea estates need to provide housing and sanitary toilets to workers. The various reports reveal that workers' house in a state of "awful poor condition" and in some cases, "cesspits overflowing into the living areas of people's homes". Tea workers in Assam earn ₹ 115 ($1.50) a day, significantly below the Assam minimum daily wage of ₹ 177. A combination of dismal living conditions and low wages has kept tea workers and their families malnourished and in poor health. The Director of Assam Medical College told BBC and Radio 4 "nine out of 10 patients" from tea plantations were malnourished. Their living conditions made them prone to diseases like diarrhoea, Tuberculosis (TB), anaemia, skin diseases, etc. The industry is also open to the criticism for engaging some children below the age of 15 years. Human Rights Institute of Columbia Law School found "abusive conditions" in Indian tea industry. (Bose, K., 2015).

All the garden labours are from the 'adivashi' community, which is the lifeline of tea gardens. Demographically, tea garden labour community of Assam represents around 20 per cent of the total population of the state accounting more than 45 lakhs tea garden labour population in the state and is one of the biggest contributors to the organised workforce. About 17 per cent of workers in Assam are engaged in tea industry and around 50 per cent of the total tea plantation workforce in Assam is women. In 1997, the State had 2470 tea gardens spread over 230 thousand hectares. Between 2005 and 2006, the state produced 476 million kgs of tea. Presently, the state has 39,139 tea estates in 507 thousand hectares of land. In 2013-14 the State produced 629.05

million kgs of tea with 304.40 thousand hectares of area under tea. Thus, it is clear that the tea production has increased. The industry has contributed considerably to the economy of Assam by providing employment to nearly half a million population along with contributing revenues and assistance to promote other road and rail network and service sector over a period of time. One cannot escape from the larger questions over making the tea garden labour still fall under the category of "indentured labourers" in the twenty-first century. In the last 150 years, the tea community in Assam never received drivable concentration in the so-called development process evidences by the state.

The tea industry in Barak Valley has been going through tough days over the last two decades. Labour problem, including less number of tea labourers for NREGA scheme, inadequate communication system, power crisis, increased revenue tax for Barak tea garden, cess on green leaf, increased pollution fee, less transport subsidy etc. have put Barak tea industry in a gloomy state. Moreover, the wages paid to tea labourers in Barak Valley region of Assam are lower than that of the Assam Valley and other states of India. Moreover, in table 1 daily wages paid to tea plantation workers of Assam Valley, Barak Valley and other states of India in 2024 has been shown. It is revealed that wages paid to the tea labourers in Barak Valley region of Assam is lower than any other major tea producing regions of India.

Table 1.1: Daily Wage rates in major Tea Producing States and Barak Valley in India (2024)

States	₹
Assam Valley, Assam	250
West Bengal	250
Tamil Nadu	406.80
Kerala	421.26
Karnataka	376.78
Barak Valley, Assam	228

Source: ITA and UPASI

The Wages of Tea plantation workers in Assam have been historically set through joint agreements of the employers and the labour Unions. Due to the expiry of the last Wage Agreement on 31/12/14, the Employers' Associations represented by the CCPA recently met the Assam Chah Mazdoor Sangha (ACMS) (largest representative Union of Tea workers in Assam) to discuss the matter of Wage revision of Daily Rated Tea Plantation workers in the Brahmaputra Valley. Following lingering discussions a 3-Year defrayal has been arrived at for raising the Daily Wage Rates from the pre-existing ₹ 94/- by ₹.43/- over 3 years. (ITA, 2015). However, the Daily Wages Rates up to 2024 are as follows:

Table 1.2: Revision of Wages in Assam Valley, Assam

Period	Wage Rate (₹/Day)
Jan, 2015-Dec, 2015	115
Jan, 2016-Dec, 2016	126
Jan, 2017-Dec, 2017	137
Jan, 2024-Dec, 2024	250

Source: ITA.

In 2024, as per the records available from the sample units, it has been found that the tea garden labourers in Barak Valley are working against a daily wage of ₹ 228 which is again lower than the wages of tea garden labourers of Assam Valley.

Under the PPP-model of National Rural Health Mission (NRHM), the tea-garden owners who have reached to an agreement with the Government represented by respective Deputy Commissioner, the owner is receiving ₹15 lakhs per annum to provide the medical facilities to the workers. Under the Plantation Labour Act, the planters are supposed to provide the medical facilities on their own. The tea-garden managements who are receiving supplementary amount of money allocated under the NRHM project are not aptly utilizing this amount and are sometimes adopting unethical means to guzzle up the money. It is a public issue of lack of observation and inspection by the authority at the appropriate level. (https://sites.google.com/site/swabhimanngo/)

Since independence the Government of India has been taking various measures in order to improve the lot of these down trodden people. Plantation Labour Act, 1951 was passed in order to bring some changes in their life. Under the Plantation Labour Act 1951, each tea garden should have a health centre with adequate facilities. Most tea gardens are remotely located and do not have proper connectivity to the nearest town areas. There are many provisions in the Act for the improvement of the health condition, educational level, recreational facilities, etc. for the tea plantation workers. In spite of these measures, many plantation workers of a large number of tea gardens have not yet been improved at a satisfactory level. At this juncture, a study to investigate the position of health, hygiene, educational level and welfare of tea plantation workers in the Barak Valley of Assam has been initiated along with examining the extent of implementation of provisions of the Plantation Labour Act, 1951, in the sample tea gardens.

1.6.Objectives of the Study:

The present study is an attempt to examine the health, hygiene, sanitation, education and welfare measures of the tea garden labourers in general along with level of implementation of relevant regulatory provisions of the Plantation Labour Act, 1951. The present study proposes to achieve the following objectives in particular.

1. To study the health, hygiene and sanitation facilities available for the tea plantation workers.

2. To investigate literacy and educational level of the tea garden workers.

3. To inspect the level of implementation of different provisions relating to health, education and welfare in the sample tea gardens.

4. To spot and identify the problems which are impeding the health and educational scenario of the tea gardens labourers.

1.7.Research Methodology

a) **Sample Frame:** In Barak Valley, there are three districts, viz., Cachar, Karimganj and Hailakandi. Most of the tea gardens are located in Cachar District. Therefore, four tea gardens of Cachar district, viz, Pallorbund, Dewan, Doyapore and Pathemara are purposively selected for the present study. As the socio-economic conditions of the tea plantation workers is more or less homogeneous 100 households from each tea garden have been selected at a random basis. Those households were excluded if no member of the households is working in the tea plantations or factories. The following table shows the sample selection:

Table 1.3: Sample Frame

Tea Gardens	Total Households (2024)	Sample
Pallorbund	670	100 (14.92)
Dewan	1270	100 (7.87)
Pathemara	750	100 (13.33)
Doyapore	565	100 (17.69)
Total	3255	400 (12.28)

Source: Primary Data and Official Record collected from the Sample gardens. Figures within brackets indicate the percentage of sample over the total households.

b) **Data Collection:** The present study is based on both primary and secondary data. Primary data has been collected from the individual households of the sample tea gardens. For collecting data on health facilities interviews with the managers and senior medical officers of the sample gardens have been conducted. Secondary data have been collected from different sources like Office of the Directorate of Economics and Statistics, Silchar, Assam; office of the Cachar Tea Labourers Union; Indian Tea Association, Surma Valley Branch, Silchar, Assam; Non-Government Organisations, Village Panchayat Books, Tea Garden Records, Journals, Websites and other unpublished works, etc.

c) **Variables Selected:** For examining the relative health facilities of the tea gardens ratios like doctor-population ratio, nurse population ratio, bed population ratio, etc have been computed. The crude death rate, birth rate and Infant Mortality Rate (IMR) of the tea labourers have been calculated to assess the extent of health awareness and health facilities of the tea labourers. Moreover, for calculating literacy rates children below the age of six years have been excluded.

d) Analysis of Data: Collected data have been tabulated, analysed, interpreted in a number of closely related operations according to nature of information

e) Period of the Study: The reference period for the present study is 2022-24.

f) Limitations of the Study: The major limitation of the study is that the end result of the study is based on the correctness of responses of the labourers to questions contained in the schedules. The responses are supposed to be accurate but the chances of documentation deceptive data cannot be entirely eliminated. As far as the secondary data are concerned there may be some sort of overlapping due to gap period in maintaining official records. Moreover, a few inconsistencies may arise due to lack of harmonization among various governmental organisations maintaining correlated and comparable records. Problems also faced due to the reluctance of the officials to provide relevant data. Also due to lack of cost and time, a small sample number of samples have been selected. The small sample may not a true representative of population.

1.8.Profile of the Study Area

The Barak valley is the southernmost part of Assam covers an extensive area of 6922 sq. kms in 3 sub-divisions of three districts viz, Cachar, Karimganj and Hailakandi. It is bounded on the north by the district of North Cachar Hills and Meghalaya state and west by Tripura and Bangladesh, on the east by Manipur and Mizoram and on the south-east by Mizoram. The economy of the valley had been a periphery to the economic mainstream of Bengal during the British rule. The partition of the country in the wake of independence and the consequent emergence of East Pakistan (now Bangladesh) had not only shattered the traditional cheapest and quickest lines of transport and communication but also snapped the age-old channels of trade, commerce and transactions. The post-independence phase did not adequately compensate the valley for all the loss inflicted on it by partition. The background of tea industries in Barak Valley revealed that in Assam tea plantation was the first capitalist enterprise introduced by the British government in the middle of the nineteenth century. Robert Bruce and his brother Charles Alexander Bruce were the pioneers of developing tea plantations in Assam. In 1839, Government started a tea garden at Chabua (near Dibrugarh) and another tea garden near Jaipur (Cachar) on experimental basis with China tea plants and seeds. In Barak Valley, the first tea garden was set up at Barshangun (present Kathal Bagan) in 1856 by G. Williamson (Verner, 1858) after the discovery of tea plants in Cachar District in 1855. Thereafter, the number of tea garden gradually increased. In 1901, the total number of tea gardens in Barak Valley was 180 (Bhattacharjee, 1977). As per statistical report, in 2001 the number increased to 206 with 32,272 hectares area under plantation. In 2005, the number of tea gardens in Barak Valley remained the same i.e. 206 (Tea Statistics, 2005). Again, in March 2009, the number of tea gardens in the valley, excluding the sick gardens, was 130. The total number of labourers employed in Barak Valley region was 60,000 permanent labourers and around 50,000 casual labourers (CCSU, 2024). Table 4 shows district wise number of tea gardens / estates operating in Barak valley of Assam.

Table-1.4: Number of Tea Gardens in Barak Valley

No. of Tea Gardens	**Barak Valley**			
	Cachar	Karimganj	Hailakandi	Total
As on 2009	72	41	17	130
As on 2013	56	24	19	99
As on 2024	56	23	19	98

Source: Indian Tea Association, Surma Valley Branch, Silchar, Assam

It is revealed from the table-4 that the number of tea gardens reduced over the years which can be attributed to liberalised trade policies adopted by the Government of India, availability of foreign cheap tea, increased cost of production, poor tea garden management, etc. The valley is economically and industrially backward. Thus, development of tea industry is urgently needed in the direction of economic development for overall development of the region under study.

1.9.Tentative Chapter Planning: the proposed study will be subdivided into following chapters:

Chapter I: Introduction
Chapter II: Health Scenario of Tea Plantation Workers
Chapter III: Educational Scenario of Tea Plantation workers
Chapter IV: Level of Implementation of Welfare Provisions in the Sample Tea Gardens.
Chapter V: Summary and Policy Recommendations
Annexure A: Bibliography

****

Chapter II : Health Scenario of Tea Plantation Workers

Health of the workers is central point for a blooming workplace, affecting productivity positively, better morale, and overall success of industry. Healthy workers are more productive, less prone to sickness and experience better job satisfaction. An emphasis on workers' health also reduces stress, improves morale, and can even augment a company's reputation. Giving priority to the health aspects not only is a matter of social responsibility, but also a shrewd business decision that can lead to augmented productivity, condensed costs, and a more prosperous workplace.
According to the International Labour Organisation (ILO), decent work is extremely vital, to assure not only the health of employees but also a sustainable, innovative and productive economic activity. The occupational health is generally well developed in Western countries, it is far less common in developing countries due to the political and economic encounters. However, the provision of occupational health has massive possibility to cut down inequality, defend vulnerable groups (such as immigrant workers, different ethnicities), and support equal access to healthcare.
A brief overview of the importance of health of the workers in an industry can be understood from the following points:

1. Upgraded Job Satisfaction: An attention on health and well-being can promote a positive work environment and superior job fulfillment.

2. Enhanced Productivity: Workers with good health are in a better condition to concentrate in the work place and complete the tasks efficiently, leading to higher output.

3. Decreased Absenteeism: Less sick days refers to less interruption to workflow and a steadier workforce.

4. Less Stress and Burnout: Wellness programs along with a sympathetic work environment can assist workers in managing trauma and preventing burnout.

5. Better Work-Life Balance: Workers with better health are found to be in a better position in managing their time and tasks, leading to a superior work-life balance.

Poor health of the workers results in reduced productivity, higher unsafe acts, and increased absenteeism. A healthy worker, on the other hand, produces results opposite to these. In other words, healthy employees are more productive, more safety conscious and are more regular to work. The worker who is healthy is always cheerful, confident looking and is an individual asset to an organization. Therefore, in the present chapter attempts have been made to examine whether the provisions relating to the health as given in 'The Plantation Labour Act, 1951' are implemented or not in the tea gardens under study and also to focus light on the certain indicators of health.

Provisions of Health and Sanitation as Mentioned in The Plantation Labour Act, 1951:

Drinking Water-

Under section 8 of the Plantation Labour Act, 1951, in every plantation effective arrangement shall be made by the employer to provide and maintain at convenient places in the plantation a sufficient supply of wholesome drinking water for all workers.

Conservancy-

As per section 9(1) of the Plantation Labour Act, 1951, there shall be provided separately for male and females in every plantation a sufficient number of latrines and urinals of prescribed types so situated as to be convenient and accessible to workers employed therein. Also section 9(2) of the Act states that all latrines and urinals provided under sub-section (1) shall be maintained in a clean and sanitary condition

Medical Facilities-

As per Sec 10-

(1) In every plantation there shall be provided and maintained so as to readily available such medical facilities for the workers [and their families] as may be prescribed by the State Government.

(2) If in any plantation medical facilities are not provided and maintained as required by sub-section (1) the chief inspector may cause to be provided and maintained therein such medical facilities, and recover the cost thereof from the defaulting employer.

(3) For the purpose of such recovery the chief inspector may certify the costs to be recovered to the collector, who may recover the amount as an arrear of land revenue.

Sub-Rr. (6) of Rule 36 prescribed u/s 10 provides that medical

and auxiliary personnel shall be appointed according to the following scale-

Qualified medical practitioner- One per every 1,750 workers or part thereof.

Midwife - One per every 1,750 workers or part thereof.

Trained Nursing attendants - One per every 300 workers or part thereof.

Pharmacist One per every 1,750 workers or part thereof

Health Assiatant One per every 2,100 workers thereof.

It has also been mentioned that a minimum of 15 beds shall be provided in every garden hospital per 1000 workers served.... (Sub.Rr. 7 (a) of Rule 36 prescribed u/s 10).

Moreover, as prescribed by the Assam Plantation Labour Rules, 1956- the im-patients are to be given diet and medicine free of cost. The diet should include one breakfast and two full meals. Breakfast's menu is to consist of tea and roti, and the meals rice, pulses, vegetable curry and an item with fish/meat/egg. In special cases milk is to be supplied

Level of Implementation of Provisions relating to Health in the Gardens under study-

With regard to drinking water, in plantation area i.e. in the area of plantations the workers have to drink water from the wells, the water of which is not hygienic as well and in this context the garden authority is not taking any proper step for purifying the water of the wells. Also, the workers are provided hot tea liquor mixed with salt to quench their thirst, while they are in work.

Through survey it has been found that none of the tea gardens is having any prescribed type of latrines and urinals in the area of plantations As a consequence the labourers are to defecate and piss in the open space and drains in the plantations. This ultimately contaminates the environment in the plantations.

With regard to medical facilities, all the four tea gardens included in the sample, have hospitals. These hospitals were set up around 70-80 years back. Let us now examine whether the tea garden management has appointed staff as per the norms laid down in Act. Table 2.1 reveals the fact

Table 2.1: Medical &Auxiliary Personnel as per the Requirement of the Act and Existing in the Tea Gardens under Study

Tea Garden	Total Workers (2023-24)	No. of Doctors		No of Pharmacists		No of Mid Wife		No of Health Assistant		No of Trained Nursing Attendant		Number of Beds	
		As per the Act	Existing	As per the Act	Existing	As per the Act	Existing	As per the Act	Existing	As per the Act	Existing	As per the Act	Existing
Dewan	2279	2	1	2	1	2	1	2	1	8	1	33	30
Pallorbund	899	1	1	1	1	1	2	1	1	3	Nil	13	14
Pathemara	1658	1	1	1	1	1	2	1	1	5	Nil	22	22
Doyapore	910	1	1	1	1	1	1	1	1	4	Nil	14	20
Total	5746	5	4	5	4	5	6	5	4	20	1	82	86

Source: Primary Data.

Thus, from above it is found that the medical and auxiliary personnel in the four tea gardens are not according to the scale mentioned in the Act. Also, the ratio of number of beds in the garden hospitals to the number of workers served is also not as per the requirements of the provisions of the Act. Moreover, in all the tea gardens included in the study, every year one eye camp is organized by NGOs to benefit the labourers and other inhabitants of the tea gardens. Also pulse polio immunization is also held in all the tea gardens. Also, all the four garden hospitals have been identified as DOTS centre for the treatment of the TB patients.

While conducting survey of the four tea gardens included in the sample, the researcher has identified the following inadequacies in the garden hospitals:

(1) There is no separate examination room

(2) There is no full -fledged labour room worth the name

(3) There is no table for minor operation worth the name

(4) Though it has been reported by the garden hospital staffs that there are separate maternity wards but in reality the mothers or the prospective mothers are kept together with other patients

(5) Poor sanitary arrangements

(6) Also in the garden hospitals, they do not have the enough stock of prescribed medicines nor do they have the prescribed instruments.

(7) Although the in-patients are given meals free of cost but only two time meal se, lunch and dinner. Breakfast is not provided. Meals include only rice and vegetable curry/pulses. In other words, fish/ meat/egg is not provided

The Status of Health and Sanitation Among the Tea Labourers.

The tea labourers are to work in the open fields ie in the tea gardens, facing scorching heat of the sun and down pouring rain and exposed to poisonous insects and snakes. In early morning they reach the duty spot i.e. the garden and the factory and work for eight hours a day with a break of half an hour for rest. But in practice, the garden workers rarely go for lunch. They leave their work place when the task is finished. There are various types of occupation in the in the tea gardens e.g. cleaning, spraying insecticides, plucking, pruning, hoeing, manuring, planting, weeding. repairing of houses, building houses and new constructions. Some of these tasks adversely affect the health of the workers.

There are more than 32 different caste and tribe groups that compose the labour population of tea gardens of Cachar district. (Das, 2006) Although they share a common economy and living, in many spheres of life, their beliefs and practices and worldview vary from one another. Nevertheless, certain problems, for example, problems of health and sanitation may be identified which they share in common.

In all the four tea gardens viz, Pallorbund, Dewan, Doyapore, and Pathemara, most of the labourers from their appearance do not look healthy. They appear to be weak, lean and emaciated. Their health conditions point to the prevalence of a general undernourishment and malnourishment due to their poor economic health in general. Consequently, they are usually found suffering from various diseases. (Debnath, 2000).
60

Tea labourers have been found to suffer from a large number of diseases like malaria, tuberculosis, diarrhoea and chronic weaknesses. Other discases like blindness, paralysis, epilepsy and skin diseases are also visible here and there in the tea gardens. Following table highlights the number of tea labourers suffering from various diseases in the four tea gardens during the period of data collection.

Table 2.2: Number of Labourers suffering from various diseases

Gardens	Tuberculosis	Diarrhoea	Anaemia/Chronic weaknesses	Malaria	Total
Dewan	3	6	21	10	40

Pallorbund	2	5	15	9	29
Pathemara	2	7	13	8	30
Doyapore	1	5	17	4	27

It is revealed from the table 2.2 that most the tea garden labourers are suffering from anaemia/chronic weaknesses followed malaria and diarrhoea. These diseases can mostly be attributed to the lifestyles and traditional practices of recovering from health problems. Moreover, most of the labourers are unaware about the hygienic lifestyles and other ways to remain healthy. Since the condition of health in a community depends upon the factors like housing, sanitation, food habit, rearing of children and individual and social awareness about health and hygiene, an analysis of these factors will reveal the logic of the situation of general ill health of the labourers in tea gardens.

Housing: Housing conditions of all the tea gardens are not same Again the housing conditions differ from garden to garden. Also housing conditions reflect socio-economic levels of the tea labourers. The conditions are shown garden wise in the table 2.3.

Table2.3: Distribution of the labourers by the type of houses

House Type	Dewan	Pallorbund	Pathemara	Doyapore
Assam Type	19	21	20	7
Bamboo Structure with Plastic Coverings/Hay	29	31	38	50
Mud Houses (Mud walls and Roof with Hay)	48	40	40	40
RCC Type	4	8	2	3
Total	100	100	100	100

Source: Primary Data.

The tea labourers in the study area are found to have three types of houses or Selter, viz, Assam Type, Bamboo Structure, Mud houses, RCC, etc. The garden authority mostly owns these houses. These houses are generally one room/two room sets. As one room cannot provide sufficient space for the workers, the workers generally make one extra room attached with the main room.

The majority of the tea garden labourers are living in the Bamboo Structure or Mud houses where they could hardly accommodate themselves. As a consequence, they have to live in a limited space and congested environment, which ultimately affects their health. In view of limited space in the houses, most of the dwellers have made small huts used as kitchens. The labourers very often keep domestic animals either in their verandas of houses or build small sheds attached to the houses. This again adversely affects their health.

The provisions as mentioned in the Plantation Labour Act, 1951 with regard to housing are-

According to section 15 of the Act, the employer in the tea gardens has to take up the responsibility of providing and maintaining houses for the workers and his family. But in reality, this is not so.

Though the garden authorities provided housing accommodation to its workers but over the last four or five years the workers are not getting any maintenance of the houses from the garden authority.

Drainage

The tea gardens included in the study are lacking proper drainage systems. The drains could hardly be seen in front of the houses of the tea labourers. In some houses, drains made by the labourers could be seen. Absence of proper drainage arrangements has converted many labour lines into pigsty with accumulation of filth and muck. This results in to the incidence of a number of diseases among the labourers. The situation worsens during the rainy season.

Water source

Wells, taps and river are the main sources of water for all purposes like drinking, bathing, washing, cooking in the garden. The labourers do not purify the water properly. The pharmacists of all the tea garden feel that poor quality of drinking water frequently causes diarrhoea among the labourers. But the labourers as they said are not aware of the causes of such diseases.

Also, the garden authorities have set up a number of wells and water supply centres are there in the gardens for the supply of water but the garden authority is not taking proper steps for maintenance of the water. In Pallorbund and Dewan though PHE has set up water supply centre for the supply of wholesome drinking water but the water supplied in most of the cases is mud mixed. Moreover, due to lack of awareness among the garden labourers, they are mostly affected by various waterborne diseases.

Hygienic Habits

Owing to their conditions and attitudes towards life, the labourers are not habitual of cleaning teeth, bathing and washing clothes daily and regularly. They clear their teeth with charcoal powder or a twig of neem tree or of a tea bush and only a few of labourers are familiar with the tooth paste. They do not bath or wash their clothes regularly but do according to their convenience or availability of water. Most of the labourers defecate in the open space, which is a cause of intestinal diseases in the garden. Generally, they defecate beside the road or the bush weeds thereby. The remaining labourers also do not have proper latrines and urinals. Moreover, toilets that are constructed in such a way as not to require daily cleaning. Thus, both defecating in the open space and the kachha toilets contributed more or less, in hygienic conditions in the tea gardens. In the area of the plantations also there is not even a single toilet or latrine for the use of the labourers. As a consequence, they have to defecate in the drains of the plantation or even in the area of plantations.

Dietary Quality

Malnutrition is the common factor affecting the general physique of the villagers because their diets do not contain sufficient quantity of nutrition. Use of eggs, milk, meat, fish and fruits is very unusual among the villagers. The habit of liquor consumption is also indirectly responsible for their undernourishment and malnutrition.

Superstitious Beliefs

Despite the availability of the medical facilities, the labourers have been depending on the superstitions for the cure of the diseases. The labourers of the tea gardens are still afraid of visiting the Sickmaan (Hospital) and like to escape from injection and saline. Sometimes the indoor patients run away from the hospital due to the fear of *Sui* (injection) and saline. Generally, the tea labourers believe that some Pesachi Deota (evil spirit) causes the diseases. For the treatment they invoke the super natural powers rather than to approach the hospital. Specially in case of pox, they will never try the modern treatment methods. But they will perform Puja (worship) to invoke and propitiate the deity who has cursed the sufferer with their strong super natural power. Whenever a person has a pain in body or suffer from paralysis, one suspects of some mischief by an enemy having invoked some supernatural forces against him. For overcoming such diseases, the labourers generally resort to an Ojha. The labourers believe that an Ojha with his super natural powers can drive out such evil spirits, which have caused such diseases. Even some of the educated labourers of the tea gardens have faith on such supernatural activities and they perform superstitions along with modern means of treatment. (Kar, 1997).

Briefly poor conditions of housing, drainage, drinking water. in hygienic habits and poor dietary quality are responsible for the present poor health among the tea labourers in the gardens under study. Moreover, the conditions of the garden hospitals should be improved to overcome the health problems of the tea labourers. In this regard the garden authority should take necessary steps otherwise poor health conditions will lead to absenteeism in the gardens, which will adversely affect the productivity of the tea gardens.

Health Indicators in the Tea Gardens:

(a) Crude Death Rate: This is considered a fair indicator of comparative health of the people It is defined as the number of deaths per 1000 population per year in a given community. Although not a perfect measure of health status, a decrease in death rate provides a good tool for assessing the overall health improvement in a population. Reducing the number of deaths in the population is an obvious goal of medicine and health care, and a success or failure to do so is a measure of a nation's commitment to better health. (Park, 2007).

Table 2.4: Crude Death Rate in the Sample Tea Gardens

Garden	Crude Death Rate (2023-24)
Dewan	9
Pallorbund	10
Pathemara	9
Doyapore	13
Overall	10

Source: Primary Data.

It is revealed from the above table that among the sample tea gardens highest crude death rate is evidenced from the Doyapore Tea Garden followed by other sample units. At this juncture, it is worthwhile to do a comparative study on the overall crude death rate of sample units with that of the Cachar, Assam and India. Table 2.5 highlights the scenario in this context.

Table 2.5: Comparative Picture of Crude Death Rate

In	Crude Death Rate
India	9.1 (2022)

Assam	6.2 (2020)
Cachar	6.8 (2020)
Sample Units	10

Source: Primary Data and Statistical Handbook Assam.

It is evident from the data in table 2.5 that the crude death rate is higher in the sample units as compared to India, Assam and Cachar district. This refers to the fact that the health facilities are not up to the required standards.

(b) Infant Mortality Rate (IMR): Infant mortality rate refers to the rate at which the death of young children under the age of 12 months or 1 year. It is the death of an infant dying under 1 year of age given a specific time period. This reveals the likelihood of the rate of death of young children under one year of age per 1000 live births. IMR is a sensitive indicator of availability, utilisation and effectiveness of health care, particularly peri-natal. (Park, 2007). In the following table, the Infant Mortality Rate in the sample tea garden is presented.

Table 2.6: IMR (per 1000 live births) in Sample Gardens

Tea Gardens	IMR (2023-24)
Dewan	6
Pallorbund	17
Pathemara	18
Doyapore	15
Overall	14

Source: Primary Data.

It can be observed from Table 2.6 that IMR is highest in Pathemara Tea Garden followed by Pallorbund, Doyapore and Dewan. Less number of IMR in tea gardens shows a better picture of health facilities in the garden hospitals as well as improved socio-economic conditions of tea labourers. Now let us compare the IMR of the tea gardens under study with that Cachar, Assam and India. Table 2.7 shows the phenomena.

Table 2.7: Comparison of Infant Mortality Rates (IMR)

In	IMR
India	26.61 (2023)
Assam	36 (2020)
Cachar	119 (2023)
Sample Units	14 (2023-24)

Source: Statistical Handbook, Assam.

It is revealed from the table that IMR is lowest in sample tea gardens. IMR is found to be highest in Cachar district followed by Assam and India.

A significant reflection evidenced during the survey with regard to labourers is that they are reluctant to go to hospital for delivery. The traditional belief of many plantation labourers is that the placenta is to be disposed of by some specific kin member within the household. So, hospitalization of a parturient mother is in many cases the last resort. On the other hand, the members of the staff from plantation hospital are reluctant to attend a patient in the residence of labourers. Thus, majority of the infant deaths is caused by pneumonia, diarrhea and nutritional deficiency. This happens mainly because of the traditional beliefs of the labourers, like- infant and the mother are to sleep on the floor, the mother is required to wash her clothes, etc. In most of the cases it has been found that mother has to drink domestically prepared wine (*popularly known as Haria*) to regain her health which may not be scientifically healthy and very often worsens the health of the mother and the new born. Usually, they come to hospital when the traditional treatment fails.

(c) Comparative Picture of Health Facilities: Following ratios have been computed to identify the comparative picture of health facilities:

- Doctor – Population Ratio.
- Nurse – Population Ratio
- Bed – Population Ratio.
- Population per Health Centre/Sub centre.

Table 2.8: Comparative picture of Health Facilities

Health Indicators	Sample Tea Gardens	Cachar	Assam
Doctor – Population	1: 5243	1:8411	1:1216
Nurse – Population	1:3677	1:3038	1:1793
Bed – Population	1:170	1:1626	1:1652
Population per Health Centre/Sub centre	1:5501	1:6539	1:6753

Source: Primary Data and Office of the Director of Health Services, Assam.

Thus, from table 2.8 we find that in case of the doctor population ratio, tea garden labourers are better than the general population of Cachar but lagging behind the general people of Assam. In case of nurse-population ratio sample tea gardens are worse off than Cachar and Assam. Again, in terms of Bed – Population ratio, the tea garden labourers of sample tea garden are better than general population. In terms of population per health centre/sub-centre we find that the tea garden labourers are in advantageous position than the people of Cachar district and Assam. In this way, it can be inferred that the health facilities of tea garden labourers are not at par when compared with Assam and Cachar as whole.

Thu, from the foregoing discussion it is clear that the problems of development health condition of tea labourers need a two-way approach for finding a solution. Alongside proper implementation of the provisions for health and sanitation facilities, the people need to be carefully motivated for availing these facilities in time. Also, tea garden labourers must be properly educated to be aware of the benefits of the modern health facilities.

******

Chapter III : Educational Scenario of Tea Plantation workers

The social thinkers have stressed the role of education in the development of individuals in all ages. Education helps in personality development, behavioural changes and skill formation of an individual. At the social level the role is very significant in the field of human resource development, social change and modernization. (Chakraborty, 1997)

In a dynamic society there cannot be one but different aims of education. These aims of education again must be related to one another. Development of individuality and social efficiency are the two broad aims of education and all other aims of education must be correlated to these broader aims.

In the earlier times the seniors in the family carried out the education process. But in the modern times separate educational institutions like schools, colleges and universities have taken the responsibility (Debnath, 2000). The educational institutions, rate of literacy and educational level are here being taken up for analysis of the role of education in the life of labourers of four tea gardens viz., Pallorbund Tea garden. Dewan tea garden. Pathemara Tea garden and Doyapore Tea garden of Cachar district. Moreover, attempts have also been made to highlight how far the provisions relating to the educational facilities of the plantation labourers as mentioned in the Plantation Labour Act, 1951 are being implemented in the tea gardens under study

Provisions relating to Educational Facilities as mentioned in The Plantation Labour Act, 1951:

Now let us see the provisions of the Plantation Labour Act, 1951 with regard to educational facilities in the plantations:

Sec 14 of the Plantation Labour Act, 1951 provides where the children between the ages of six and twelve of workers employed in any plantation exceed twenty-five in number, the State Government may make rules, requiring every employer to provide educational facilities for the children in such manner and of such standard as may be prescribed.

Rule 52 prescribed under section 14 provides that an employer may not provide and maintain a Primary school if there is one under the direct management of the State Government or of any local body for imparting free education to the children up to the primary or higher standard, with enough seat to admit the children between the ages of six and twelve of the workers in his plantation and within a distance of one mile from the place where workers reside in the plantation, or if under any law he is required to pay a cess or tax for primary education,

Provided further that to the provisions of sub-R. (1) Of R 53 a group of employer may jointly provide and maintain a Primary School and share its expenses.

Rule 53 provides that

(1) Every school to be provided and maintained under R.52 shall be centrally situated.

(2) The school building shall be constructed in accordance with the standard plan or plans, which may be laid down by the State Education Department

Provided that where there is no standard plan or where it is proposed to deviate from a standard plan, the State Education Department's approval of the plan of the school building shall be obtained.

(3) Where adequate place is available, an open-air playground with suitable accessories shall also be provided for the children attending the school.

Rule 54 states that the employer or employers, as the case may be. shall provide for every Primary School maintained under R. 53 such educational and other requirements as may be considered necessary by the State Education Department

Rule 55 provides that -

(1) The employer or employers concerned shall appoint one teacher for every forty children or part thereof on the roll of the Primary School.

(2) The teacher shall possess the qualification prescribed by the State Education Department for teachers in Government Primary Schools:

Provided that in the case of any person who is working as a school teacher in a plantation at the commencement of these rules the State Education Department may, subject to such conditions as it may specify, relax any of the qualifications.

Level of implementation of these provisions:

Thus, it is revealed from the above-mentioned provisions of the Plantation Labour Act, 1951 that the tea garden authority has to take the responsibility to establish and run primary schools in the gardens if there is no such school under the control and management of the State Government. But in all the tea gardens, included in the sample, the garden authority established schools before the enactment of the Act.

The employees of the tea gardens were the teachers in these schools The garden authority provided books free of cost to the students. But after independence, gradually the State Government took over these schools. High Schools of Dewan and Pallorbund tea garden are taken over by the State Government and the garden authority is not managing these schools anymore. Earlier the salaries of the high school teachers were paid by the garden authority.

Educational Institutions:

The educational structure existing in the institutions of the state of Assam consists of three levels, viz., Primary School up to fourth standard, Middle School education from fifth to seventh, High

School from eighth to tenth, Secondary from eleventh to twelfth standard and three year degree course of Graduation and then Post Graduation Both State sponsored and privately sponsored schools and colleges are there in Assam Detailed account of the educational institutions of the four tea gardens included in the sample are given below:

Pallorbund Tea Garden:

There were one LP. School, one M.E. School and one High School in the tea garden area of Pallorbund. Pallorbund Garden L. P. School established in 1950 (provincialized in 1974) provides education up to 4th standard and the one is the Pallorbund M. E school established in 1966 (provincialized in 1978) providing education up to 7th standard. Pallorbund Garden Public High School established in 1993 (Provincialized in 2012) provides education up to secondary school certificate (10 standard). After high school level education a student has to go to Nena Meah Higher Secondary School, Banskandi situated 5 km away from the garden.

Dewan Tea Garden:

In this garden also there is one Primary school- Dewan LP. School established in 1931(provincialised in the later part of 1960s) provides education up to 4 standard and the Pearson Institute of M. E. School established in 1948 (provincialised in 1970) provides education up to 7th standard. Dewan Group High School established in 1982 provides education up to secondary school certificate (10 standard). Then, there is a Nehru Higher Secondary School in Poilapool which imparts education up to H. S School certificate (12th standard). Therefore, after high school level education, a student has to Nehru Higher Secondary School

Pathemara Tea Garden:

In this garden also there are three schools. Students belonging to the garden generally go to Pathemara Garden LP. School, established in the year 1936 (provincialised in 1995) which provides education up to 4th standard and Khaspur Krishnasadhan Burman M. E. School, established in the year 1962 (provincialised in 1977) provides education up to 7th standard. Krishnasadhan Burman Janaki Roy Memorial High School, established in 1969 (provincialised in 1982) provides education up to 10th standard. After completion of the H.S.L.C. level the students for further education have to take admission in D.N. Higher Secondary School situated at Udharbund 10 km away from Pathemara tea garden, which provides education up to 12th standard.

Doyapore Tea Garden:

In Doyapore tea garden, there is only one LP school- Doyapore Garden L P School established in the year 1936 (provincialised in 1972), which provides education up to 4th standard. After completion of the primary level the pupils generally go to Happy Valley ME School situated at Majhagram tea garden, 3 km away from the Doyapore tea garden which provides education up to 7 standard and after that the students have to go to D. N. Higher Secondary School, situated at Udharbund, 4 km away from Doyapore which provides education up to 12th standard.

Moreover, under Sarba Siksha Abhiyan (SSA) in all the tea gardens additional 2/3 buildings had been constructed alongside the garden LP School building, since 2002-2003. Also, books are provided to the students free of cost up to 7th standard. Under Mid-day Meal Programme students are provided lunch. Under SSA grant different schools have been started in different lanes and sub lanes of the tea gardens for providing free primary education to the wards of the labourers. These schools (under SSA grant) are run under the control of Head Teachers of the LP Schools of the gardens. The Head Teachers are of the opinion that in spite of these facilities the labourers are not sending their progenies regularly to the schools. Sometimes the teachers have to take the initiative for bringing the students in to the school. These schools are now also providing education to children residing near the tea garden area as well. This points to the lack of awareness of tea plantation workers towards the beneficial effects of education.

Rate of Literacy and the level of Education among the Labourers in the Gardens under Study:

In spite of a network of schools spread in and around the tea gardens included in the study the literacy and educational level of the garden workers are low. The rate of literacy of the sample households is depicted in the table no 3.1.

Table 3.1: Sex wise Literacy Rate of the Tea Labourers

Gardens	Literate			Illiterate			% of Literacy		
	Male	Female	Total	Male	Female	Total	Male	Female	Total
Dewan	211	105	316	91	102	193	69.86	50.72	62.08
Pallorbund	219	111	330	125	76	201	63.66	59.35	62.15
Pathemara	229	122	351	114	81	195	66.76	60.10	64.29
Doyapore	199	130	329	101	88	189	66.33	59.63	63.51
Total	858	468	1326	431	347	778	66.56	57.42	63.02

Source: Primary Data.

In table 3.1, the garden wise literacy rate of the labourers is shown. For this purpose, population in the age group of 0-5 years is excluded. Out of the sample tea gardens, literacy rate is found to be the highest in Doyapore tea garden (63.51%). Now, if we look at sex wise, literacy rate is higher in case of males as compared to females. The wide prevalence of illiteracy among the women points to the existence of conventional attitude towards the women education. Again, the percentage male literacy is highest in Dewan Tea Garden and highest female literacy is found in Pathemara Tea Garden. The lowest percentage of male literacy is found in Pallorbund Tea Garden and also the lowest female literacy is found in Pallorbund tea garden. The overall literacy rate in sample tea garden is found to be 63.02%. Now, let us compare the literacy rate in the sample tea gardens with that of Cachar, Assam and India in table 3.2.

Table 3.2: Comparative Picture of Literacy Rates

People of	Literacy Rate (in Percent)		
	Male	Female	Overall
Sample Tea Gardens (Overall)	66.56	57.42	63.02
Cachar	84.78	73.68	79.34
Assam	77.85	66.27	72.19
India	82.14	65.46	74.04

Source: Primary Data and Census Report.

From the above table, it is revealed that overall literacy rate of the sample tea garden labourers is lower than that of the overall literacy rates of Cachar, Assam and India. If we compare the sex wise literacy rates, the sample tea gardens are still lagging behind the district, state and national literacy rates. There also exists a wide gap between the literacy rate of male and females. So, proper policy and guidelines must be developed at the appropriate level to address the issue and steps must be initiated to improve the literacy rates in the tea gardens.

At this juncture it is worthwhile to investigate the educational levels of the labourers in the sample tea gardens. So, in table 3.3, the educational level of tea labourers has been shown.

Table 3.3: Educational Level of Tea Plantation Workers

Level of Education	Dewan	Pallorbund	Pathemara	Doyapore
Primary Level	115 (36.39)	136 (41.21)	157 (44.73)	164 (49.85)
Middle School	159 (50.31)	165 (50.00)	159 (45.30)	132 (40.12)
High School	23 (7.28)	21(6.36)	27 (7.69)	16 (4.86)
H.S. Level	9 (2.85)	6 (1.81)	6 (1.71)	9 (2.74)
Graduation	2 (0.63)	2 (0.60)	1(0.28)	8 (2.43)
Post-Graduation	2 (0.63)	0	1 (0.28)	0
Total	316 (100)	330	351	329

Source: Primary Data. Figures within parenthesis indicate percentage to total.

The data in table 3.3 indicate the educational level of tea labourers. It is found that most of the labourers are educated up to the primary or middle school level. Very less number labourers are reaching up to the H.S. level or graduation or post-graduation level. This also brings out the lack of institutions providing education up to graduation or post-graduation. There were no colleges situated nearby the garden area. Though most of the labourers are first generation learners but still they are having the ambition of higher education for their children.

It has been evidenced from the survey that the tendency of the labourers to send their children to school has increased over the years in expectation of a better life. Generally, the children in the garden get admitted into a school a bit late. So, in the primary level students between 7 to 10 years are mostly found. The attendance of the students is more or less regular. The headmaster of the primary schools informed that the attendance of students in school is near about 75%. Dropout of students from schools is regular feature as the parents of children are always not able to spare them for education. The parents consider their children as economic assets as soon as they reach at the age of 10 to 14 years. These children assist their family by doing job as minor labourers at the garden, collecting firewood, and looking after younger brother or sister.

Communication: - Media and their uses

The garden labourers communicate with each other by personal contacts only. They use both personal as well as mass media for communicating with the outside world. The tea gardens included in the study are connected with other gardens and Silchar by pitch roads. The garden labourers go by cycles or on foot to deliver a message in the neighbouring gardens. Some of the literate villagers have also access to newspapers to expose them to the dynamics and happenings in their im(mediate as well as far off situations. Their habits of reading newspapers will reveal their attitude towards this type of media. The young generations among the tea labourers frequently watch televisions, view cinemas or watch android mobiles. Many of the labourers have DTH (like-Airtel Dish TV, Tata Play, Dish TV, etc.). some of them are found to use social media accounts like- whatsapp, facebook, etc. Some are found using keypad mobiles. Most of the labourers are found to use mobile phones mostly for communication purposes. This phenomenon can be understood from the Table 3.4.

Table 3.4: Households With T.V., Keypad Mobiles and Android Mobiles

Items of Possession	Dewan	Pallorbund	Pathemara	Doyapore
T.V.	67	59	63	51
Keypad Mobiles	27	28	31	33
Android Mobiles	29	31	23	16

Source: Primary Data.

The data presented in table 3.4 points to the significant prevalence of communication media. Most of the households are having a T.V. Also, a good number of labourers are found to use keypad mobiles. Some of the labourers are also found to using the android mobiles. These means of communication have wide implications with regard to exposure of the plantation labourers towards the external world. In what way this ever-increasing exposure is consequential for their development is the question to be researched out.

Reading Habit of Newspapers by the households

Identification of the reading habits of newspaper is a very vital step towards the identification of the way they utilise their competencies. The reading habits are displayed in the following table.

Table 3.5: Households by their habits of Reading Newspapers

Type of Readers	Dewan (100)	Pallorbund (100)	Pathemara (100)	Doyapore (100)	Total (400)	Percentage to Total
Regular	27	24	22	19	92	23.00
Casual	23	32	21	23	99	24.75
Total	50	56	43	42	191	47.75

Source: Primary Data.

The data show that only 47.75% of the households investigated, read some newspapers. Again 23.00% are regular readers and 24.75% are casual readers. The regular readers are subscribing papers regularly. They read Bengali local dailies, viz., Dainik Jugasankha, Dainik Samoyik Prasanga, etc. The regular readers are qualified mostly up to middle school, high school level and H. S. Level (Class XII) respectively. The casual readers generally depend on the availability of newspapers from various sources like nearby shop, quarters of the garden staffs, or political worker's house. The newspapers read by the garden labourers are in Bengali language These newspapers are published from Silchar, ie., district headquarter town.

The low profile of the habits of reading newspapers is contingent upon their educational and economic levels. The number of the newspapers read by the respondents, sources of their availability and the diversifying economy of the gardens, all have their roles in the ongoing process of habit formation for newspaper reading among the villagers

To sum up, Newspapers, TV and Radio play an important role in creating awareness among the masses on different matters like importance of education, health and eradication of superstitious beliefs. But it has been found that less than 50% of the sample households are reding newspapers.

Exposure of Tea Labourers towards Digitisation

Digitisation has led a revolutionary change in the field of education by promoting online learning platforms, virtual classrooms and personalised learning. In other words, Digitization meaningfully affects the life of a person by rationalization processes, enhancement of efficiency, and providing prompt access to information and services. It also nurtures invention, promotes alliance, and empowers individuals with latest tools and opportunities. Digitisation provides the following benefits:

- **Boosted Communication**
- **Fast Access to Information**
- **Suppleness and Convenience**
- **Improved Healthcare.**
- **Change in Personal Life**

Many of the households surveyed is found to have access to T.V. with DTH and Mobile phone. Although, most of the schools in the study area are found to be not using any digital technology in imparting the education. So, at this juncture it is useful to analyse the perception of the tea labourers towards digitisation in the field health and education. Table 3.6 reflects the same:

Table 3.6: Perception of Tea Labourers towards Digitisation

Opinion	Frequency	Weights	Product (Frequency X Weight)
Strongly Agree	59	5	295
Agree	60	4	240
Neutral	55	3	165
Disagree	136	2	272
Strongly Disagree	90	1	90
Total	400	15	1062

Source: Primary Data.

$$\text{Likert's Score} = \frac{\textbf{Total of Product}}{\textbf{Total No.of Respondents}}$$

$$= \frac{1062}{400}$$

$$= \mathbf{2.65.}$$

From the table 3.6 it is revealed that the Likert's score of 2.65 which is below the weighted average score of 3. Thus, it can be concluded that the exposure of tea labourers towards digitisation does not improve their awareness towards health and educational aspects among them. Hence, in this context proper awareness must be initiated so that the tea labourers become aware about the positive aspects of digitisation towards the improvement of their health and educational scenario.

However, the researcher has identified the following problems with regard to the education in tea gardens-

(1) Economic Problems: -

(a) The parents cannot afford the expenditure of the education of their wards. Though the primary education has been made universal and free, the stage wise admission from class I to IV, V to VII, VIII to X are not affordable to many parents and as such they are compelled to withdraw their wards from one stage to other.

(b) Children at the age group nine to ten years, actively help their parents in their productive occupation. So, in order to meet family expenses. children are withdrawn from school as soon as they attain workable age to help their parents in earning. Thus, the poverty contributes greatly to withdraw the children from the education.

(c) The physical environment and system of education in the schools of sample tea gardens may not be students friendly.

(2) Social Problems: -

(a) There are also some people among the garden labourers who are not interested in continuing the education of their children because of their social set up.

(b) Most of the parents or grandparents are illiterates, they do not comprehend the importance of education. Therefore, they withdraw their child from the school at a very early age.

(c) Practices of early marriage among tea garden workers can act as a frontier to girls' access to education.

(d) Due to shortage of Vocational Training Institute nearby tea gardens hinder the opportunities for vocational training and skill development, impacting their prospects of employment generation.

(3) Educational Problems: -

(a)Schools curriculums have been developed in isolation from the real life of the children. As a result, it becomes boring for the children. Also, teachers adopt dull and defective methods of instruction.

(b) Teachers are sometimes overloaded with some other works, which prevent them from paying individual attention to the students.

(c) Students are also sometimes scared to face the examination.

(d) The heterogeneous facilities like Mid-Day Meals, medical facilities, textbooks and writing materials to the children are not uniformly given.

(e) Lack of sufficient number of teachers also creates problems in the schools.

(f) There is also lack of proper equipment to impart education in the schools.

(g) Lastly, parents feel that education of children isolates them from the habits of work and as a result they prefer to withdraw their children from the schools.

In conclusion, the education of tea garden workers is hampered by a complex network of socio-economic, cultural, and environmental factors. Addressing these challenges needs a diversified approach, including providing access to quality education, improving living conditions, promoting awareness about the importance of education, etc.

****

Chapter IV : Level of Implementation of Welfare Provisions in the Sample Tea Gardens

Labour welfare entails all those activities of the employers, which are directed towards providing the certain facilities and services in addition to wages and salaries. Welfare services are merely maintenance factors and not motivators, yet they are essential for the health of the organization since they bear a close connection with productivities of the labour force.

Hence in the present chapter attempts have been made to study the various provisions relating to welfare of plantation labourers, relevant to health and education, as included in The Plantation Labour Act, 1951, and also to examine level of implementation of these provisions.

Provisions of Welfare as mentioned in the Plantation Labour Act, 1951:

Canteens (section 11)

(1) The State Government may make rules requiring that in every plantation wherein one hundred and fifty workers are ordinarily employed, one or more canteens shall be provided and maintained by the employer for the use of the workers.

(2) Without prejudice to the generality of the foregoing power such rules may provide for -

(3) The date by which the canteen shall be provided;

(b) The number of canteens that shall be provided and the standards in respect of construction, accommodation, furniture and other equipment of the canteen;

(c) The foodstuffs which may be served therein and the charges which may be made therefore,

(d) The constitution of a managing committee for the canteen and the representation of the workers in the management of the canteen,

(e) The delegation to the chief inspector, subject to such conditions as may be prescribed, of the power to make rules under clause (c).

Crèches (Section 12)

(1) In every plantation wherein fifty or more women workers (including women workers employed by any contractor) are employed or were employed on any day of the preceding twelve months, or where the children of women workers (including women workers employed by any contractor) is twenty or more, there shall be provided and maintained by the employer suitable rooms for the use of children of such workers.

Explanation For the purposes of this sub-section (1-A), "Children" means persons who are below the age of six years.

(1-A) Notwithstanding anything contained in sub-section (1), if, in respect of any plantation wherein less than fifty women workers (including women workers employed by any contractor) are employed or were employed on any day of the preceding twelve months, or where the number of children of such women workers is less than twenty, the State Government, having regard to number of children of such women workers deem it necessary that suitable rooms for the use of such children should be provided and maintained by the employer. it may, by order, direct the employer to provide and maintain such rooms and thereupon the employer shall be bound to comply with such direction.

(2) [The rooms referred to in sub-section (1) or sub-section 1-A] shall-

(a) Provide adequate accommodation,

(b) Be adequately lighted and ventilated,

(c) Be maintained in a clean and sanitary condition; and

(d) Be under the charge of a woman trained in the care of children and infants.

(3) The State Government may make rules prescribing the location and the standards of the rooms referred to in sub-section (1) or sub-section (1-A) in respect of their construction and accommodation and the equipment and amenities to be provided therein.

Recreational Facilities (Section 13)

The State Government may make rules requiring every employer to make provisions in his plantation for such recreational facilities for the workers and children employed therein as may be prescribed.

In addition, rule 50 prescribed under section 13 provides-

(1) Recreation centre to the scale of one for every one hundred and seventy-five (75) families of resident workers or part thereof with provision for a radio set and indoor games for adult and child workers as specified by the Chief Inspector of Plantations with prior approval of the State Government subject to there being at least one centre for each plantation or out garden thereof.

(11) Where adequate flat open space is available within a reasonable distance, a playground or playgrounds for the adult and child workers with necessary sports equipment for outdoor games

Also, every recreation centre to be provided and maintained under R 50 shall be conveniently situated as near as possible to the workers' (Rule 51)

Welfare Officers (Section 18)

(1) In every plantation wherein three hundred or more workers are ordinarily employed the employer shall employ such number of welfare officers as may be prescribed.

(2) The State Government may prescribe the duties, qualifications and conditions of service of officers employed under sub-section (1)

Other Facilities (Section 17)

The State Government may make rules requiring that in every plantation the employer shall provide the workers with such number and type of umbrellas, blankets, rain coats or other tile amenities for protection of workers from rain or cold as may be prescribed.

Rules Prescribed under section 17

Rule- 71: Facilities for protection from weather- every employer shall supply to every worker free of cost -

(a) One umbrella for every two years,

(b) One blanket or jersey for every two years,

(c) One pair of chappal for every year, and

(d) One raincoat for every two years, of the type and standard as prescribed by the chief inspector from time to time with prior approval of the State Government.

The provision of this rule shall not operate to the prejudice of any right to which a worker may be entitled to under the terms of any reward, agreement, settlement or contract of service.

Weekly Hours (Section 19)

(1) No adult worker shall be required or allowed to work on any plantation in excess of [forty-eight hours] a week and no adolescent or child for more than [twenty-seven hours] a week.

(2) Where an adult in any plantation on any day in excess of the number of hours constituting a normal working day or for more than forty-eight hours in any week, he shall, in respect of such overtime work, be entitled to twice the rates of ordinary wages:

Provided that no worker as such shall be allowed to work for more than nine hours on any day and more than fifty hours in any week

(3) For any work done on any closed holiday in the plantation or any day of rest, a worker shall be entitled to twice the rates of ordinary wages as in the case of overtime work.

Weekly Holidays (Section 20)

(1) The State Government may by rules made in this behalf-

Provide for a day of rest, in every period of seven days, which shall be allowed to all workers.

(b) Provide for the conditions subject to which, and the circumstances in which, an adult worker may be required or allowed to work overtime.

(2) Notwithstanding anything contained in clause (a) of sub-section (1) where a worker is willing to work on any day of rest, which is not a closed holiday in the plantation, nothing contained in this section shall prevent him from doing so:

Provided that in so doing a worker does not work for more than ten days consecutively without a holiday for a whole day intervening.

Explanation 1- Where on any day a worker has been prevented from working in any plantation by reason of tempest, fire, rain or other natural causes, that day, may, if he so desires, be treated as his day of rest for the relevant period of seven days within the meaning of sub-section (1)

Explanation II- Nothing contained in this section shall apply to any worker whose total period of employment including any day spent on leave, is less than six days.

As per sub-rule (2) of Rule 72, prescribed under section 20, every worker shall be paid for work on the day of rest at the overtime rate prevailing in a particular area as fixed under the Minimum Wages Act, 1948, and where there is no such rate at double the ordinary rate.

Daily Intervals for Rest (Section 21)

The period of work on each day shall be so fixed that no period shall exceed five hours and that no worker shall work for more than five hours before he has an interval for rest at least half an hour.

Spread-Over (Section 22)

The period of work of an adult in a plantation shall be so arranged that inclusive, of his intervals for rest under section 21 it shall not spread over more than twelve hours including the time spent in waiting for work on any day.

Annual Leave with Wages (Section 30)

(1) Every worker shall be allowed leave with wages for a number of days calculated at the rate of-

(a) If an adult, one day for every twenty days of work performed by him, and

(b) If a young person, one day for every fifteen days of work performed by him.

Explanation – (1) for the purposes of calculating leave under this sub-section-
(a) Any day on which no work or less than half a days work is performed shall be counted, and

(b) Any day on which half or more than half a day's work is performed shall be counted a one day

Explanation- (2) The leave admissible under this sub-section shall be exclusive of all holidays, whether occurring during, or at either end of, the period of leave.

(2) If a worker does not in any one period of twelve months take the whole of the leave owed to him under sub-section (1), and leave not taken by him shall be added to the leave to be allowed. to him under that subsection in the succeeding period of twelve months.

(3) A worker shall cease to earn any leave under this section when the earned leave due to him amounts to thirty days.

(4) If the employment of a worker who is entitled to leave under this section is terminated by the employer before he, has taken the entire leave to which he is entitled, the Employer shall, pay him the amount payable, under section 31 in respect of the leave not taken, and such payment shall be made before the expiry of second working day after such termination.

Sickness and Maternity Benefits (Section 32)

(1) Subject to any rules that may be made in this behalf. every worker shall be entitled to obtain from his employer-

In the case of sickness certified by a qualified medical practitioner, sickness allowance, and

(b) If a woman, in the case of confinement or expected confinement, maternity allowance,

At such rate, for such period and at such intervals as-may be prescribed.

(2) The State Government may make rules regulating the payment of sickness or maternity allowance and any such rules may specify the circumstances in which such allowance shall not be payable or shall cease to be payable and in framing any rules under, this section the State Government shall have due regard to medical facilities that may be provided by the, employer in any plantation.

Rule 74 prescribed under this section mentions-

(1) Every worker shall be entitled to obtain sickness allowance from his employer for each day of certified sickness for a total period of 14 days in a year at the rate of two thirds of his daily wages or the average daily earnings of the day on which he had worked during the preceding three months, if he is employed on piece rates, should be higher:

Provided, however, that where under existing arrangements the sickness allowance is higher, worker shall not, by dint of this rule, be deprived of such higher allowance:

Provided further that the worker shall not be entitled to the allowance for an initial waiting period of two days except in the case of spell of sickness following, at an interval of not more than fifteen days, the spell of sickness for which sickness allowance was last paid.

(2) Sickness allowance specified under in sub-R. (1) shall be paid at the option of the worker either at the time of payment of weekly advances on account of wages payable to the worker after he resumes his works on recovery from certified sickness.

With regard to maternity allowance, it has been mentioned in sub-R. (2) of Rule 75 prescribed under section 43, a woman shall not be entitled to obtain such allowance if she has not -

(a) permitted herself to be medically examined or the purposes of certifying pregnancy or expected confinement by a Certifying Surgeon or a qualified resident medical officer with the assistance of nurse, if the surgeon/ medical officer is male or qualified lady medical practitioner, or

(b) during the period which extends from the commencement of the four weeks immediately preceding the expected day of her delivery to the day of her delivery and during the period of eight weeks immediately following the day of her delivery attended, or permitted herself to be treated in a hospital by a qualified medical practitioner or under the supervision of such medical practitioner by a nurse:

Provided that non-compliance by the woman with the provisions of this sub-rule shall not disentitle her to maternity allowance if a certifying surgeon or any other qualified medical practitioner certifies that in his opinion such non-compliance was due to premature delivery or to a bona fide miscalculation on the part of the woman as to the state of advancement of her pregnancy.

Also, in this context Rule 76 provides-

(1) No employer shall knowingly employ a woman in a plantation during the four weeks immediately preceding the expected day of her delivery and during the eight weeks immediately following the day of her delivery.

(2) In case of miscarriage a woman shall, on production' of a certificate signed by the Certifying Surgeon or any other qualified medical practitioner or such other proof of miscarriage as may be accepted by her employer, be entitled to two weeks' leave from the day of her miscarriage with pay under the rules.

(3) During the period of pregnancy, a woman shall not be employed on work which is arduous or which requires long hours of standing at one place or may in any way interfere with her pregnancy and is likely to cause miscarriage or adversely affect health and interfere with normal development of foetus.

Also, as per the provisions of the Act the monthly ration facility is to be allowed to all the workers according to the following criteria:

14 kg per worker (@ ₹ 0.54/kg)

10 kg per dependent above the age of 8 years (@ ₹ 0.54/kg).

5 kg per dependent in the age group of 2.5 years to 8 years (@ ₹ 0.54/kg).

Level of Implementation of these Provisions:
Now let us see how far the management of the tea gardens under study has implemented the above-mentioned provisions relating to the welfare of tea plantation labourers.

1. Canteens (Section 11): None of the tea gardens under study is found to have established any canteens in the area of the plantation. While collecting the primary data it has been found that the labourers are arranging their lunch of their own. They are carrying the food from their houses to the place of work and these are kept until the time of lunch break. They prepare their lunch early in the morning. which sometimes loses its quality at the time of lunch break.

2. Crèches (Section 12): None of the tea gardens is found to have established any crèches as per the Act in the area of plantation. The women workers are either dispensing with their work or leaving their Childs in the house under the care of someone else. This sometimes adversely affects the health of the Childs.

3. Recreation Centres (Section 13): Recreation Centres are also not found in any of the tea gardens under study. Earlier there were recreation centres in the gardens. In all the four tea gardens included in the study, garden authority provided one T.V. set for the labourers and this was kept in Garden Nach Ghar (a well-constructed auditorium type place of Worshiping). The labourers used to come to the Nach Ghar for viewing cinemas, etc. But over the last twelve or thirteen years these T.V. sets are traceless. Moreover, playgrounds are there in all the four tea gardens but the garden authority is not regularly providing any sports equipment for outdoor as well as indoor games.

4. Other Facilities (Section 17): The management is also not properly implementing the provisions with regard to the facilities for protection from weather as mentioned in Rule 71. prescribed under Section 17. Over the last four or five years the labourers in the gardens under study have not been provided any umbrella, blanket or jersey, pair of chappal and raincoat.

5. Welfare Officers (Section 18): The tea gardens under study are not having any welfare officers. The concerned authority is also not taking adequate steps for the appointment of required number of welfare officers in the gardens. In fact the assistant managers who looks after the works of field (plantation area) are considered as welfare officers in the gardens. But these managers do not look after the welfare of the labourers, instead they remain busy in scheduled administrative work. At present, in Pallorbund and Doyapore tea garden, no such manager is there as well.

6. Weekly Hours (Section 19): It has been found that the workers are working more than the specified time limit. The rate of wages paid for such overtime work is not double but equal to the rate of wages paid in scheduled working conditions. This becomes possible on the part of the management, partly on account of the poor economic conditions of the labourers and partly on account of lack of awareness among the labourers.

7. Weekly Holidays (Section 20): It requires that employer must provide for a day of rest to its labourers in every period of seven days. But in practice, the garden labourers rarely get any day of rest during the season time (from April to October). Though specified holidays are there in all the tea gardens (e.g., in' Pallorbund it is Tuesday; in Dewan it is Wednesday; in Pathemara it is Thursday; in Doyapore it is Sunday) but the workers are generally forced for work by giving the allurement of cash wages in those holidays during the busy season every year (i.e., from the month of March to November). It has also been mentioned in the section that a worker should not work for more than ten days consecutively. But in reality, during the busy season time the workers are found to be working continuously in the plantations for more than thirty days. It has also been mentioned in the aforesaid section that every worker shall be paid for work on the day of rest at the overtime rate prevailing in a particular area as fixed under the Minimum Wages Act, 1948, and where there is no such rate at double the ordinary rate. It has been found the rate of wages paid to the workers during the day of rest is equal to rate of wages paid in other working days. Here also this sort of exploitation by the employers becomes possible on account of ignorance and poor economic conditions of the labourers.

8. Daily Intervals for Rest (Section 21): The provisions of this section are more or less implemented by the employers. The workers are allowed at least half an hour rest after four hours of continuous work. Sometimes the workers are reluctant to take rest, as they want to finish their assigned task before leaving for home. The daily task assigned to the garden worker is called 'Nirikh'. In plucking, the 'nirikh' is 18 to 21 pon (1 pon=1kg) per day. If any worker is plucking less than their 'nirikh' then he or she will get half of actual daily wages. That is why very often the plantation labourers are not willing to take rest.

9. Spread Over (Section 22): The provision of this section is also more or less followed by the employers of the tea gardens under study. It has been found through survey that the time is not exceeding twelve hours instead it is less than eleven hours including daily intervals for rest.

10. Annual Leave with Wages (Section 30): Annual Leave with Wages is implemented in the gardens included in the study. But variations are found with regard to provision of such facilities in the case of adolescent workers. In other words, the management is not willing to provide such facilities to the adolescent workers. The adult workers are properly availing these facilities every year. Also, with regard to wages during the leave period, some variations are found. Though the garden authority is providing the leave with wages as per the requirements of section 30 but they are not ready to provide at the rates specified in the Act. This is very much true in the case of adolescent workers. But the adult labourers are provided wages during the leave period at the rate specified in the Act.

11. Sickness and Maternity Allowance (Section 32): With regard to implementation of the provisions relating to the sickness and maternity allowance as per section 32 of the Plantation Labour Act, 1951, in the tea gardens included in the study the following discrepancies were found.

(i) The management of the tea gardens is delaying the payment of such allowances in the gardens.
(ii) The garden authority is not allowing the accumulation of annual leave with sick leave at the discretion of the workers.
(iii) They are also not willing to allow the accumulation of sick leave.
(iv) Due lack of proper communication between the garden hospital and the garden central office very often pregnant women workers are found in a plantation during four weeks immediately preceding the expected day of her delivery.
(v) Also, some pregnant women were found in arduous work in the plantations.
(vi) Very often the maternity allowances are not paid as per the rate specified in the Act.
(vii) Moreover, variations were found with regard to period of maternity leave.

Moreover, through survey it has been found in the gardens that the Casual Workers are not provided ration evenly throughout the year. Ration facility is allowed to the Casual Workers, only from the month of July to October every year. But the permanent workers are availing this facility evenly throughout the year. Again, among the permanent labourers there are two groups viz.,
(a) those who work throughout the year and
(b) those who work from the month of April to November every year.

The ration facility is allowed evenly to the first group of permanent workers and the second group is availing this facility only during the period of their work. Actually. such type of division among the permanent workers is made by the management every year due to the reduction of work assignments in the plantations in those periods every year (i.e., from the month of December to March). In this context it is to be mentioned that as per the agreement between the garden authority and the labour union the casual labourers are also to be provided ration facility evenly throughout the year.

From the foregoing discussion on the level of implementation of welfare provisions relevant to health and education as included in The Plantation Labour Act, 1951, it is revealed that the management of the tea gardens is not properly implementing the provisions. The lack of awareness about the existence of such provisions among the labourers and their level of literacy are to some extent responsible for that. Besides, the low bargaining power of the tea labourers is also acting as a barring factor in path of the proper implementation of these provisions.

****

Chapter V : Summary, Conclusion & Suggestions

The present study has been undertaken to examine the health and educational scenario along with the level of implementation of the various provisions of the Plantation Labour Act. 1951, relating to health and education in four tea gardens of Cachar district viz., Pallorbund, Dewan, Pathemara and Doyapore. Attempts have also been made to present a brief picture relating to health, sanitation and educational scenario of the plantation workers in the tea gardens under study

The uniqueness of the tea plantation industry in Cachar lies in its dualistic nature of economic structure. The industry has a formal organization. It has an elaborate industrial bureaucratic structure with a hierarchy of staff. In this hierarchy there is a manager at the top and the workers i.e. the labourers at the bottom with a group of intermediate staff in between. All the levels are having their specific duties and obligations These features of industrial bureaucracy in plantations largely correspond to that of the industrial capitalism.

Tea population in district of Cachar constitutes a large segment of the total population of the district. The tea workers in the district mainly migrated from the states of Bihar, Orissa, and West Bengal. There are more than 32 caste and sub caste in the tea gardens of Cachar district. In fact, tea garden work force is the conglomeration of the population belonging to different castes, tribes, religion and provinces but living together for more than a century as a distinct class of 'tea tribe whose contribution to the national economy is going to be sunk in the oblivion past unless proper attention is given towards the improvement of their health and educational scenario.

But the way the industry earlier recruited labourers was similar to the feudal features than to the capitalist ones. The Arkatti and the Sardari systems of recruitment did not resemble with the recruitment system practised in the capitalist enterprises. The Arkattis and the Sardars were commonly known as "coolicatchers" and their mode of recruitment as "coolie-catching". Then the conditions of those migrated labourers were very much dilapidated. They were forced to accept another contract after the expiry of the earlier one. The breach of contract was made criminally punishable. The workers were then treated like serfs, who were tied to the soil by offering plantation land. They remained perpetually bonded by borrowing money from the manager or the planters. Those features of plantations closely resembled to that of feudalism. The planters appeared as the capitalist owners of the plantations and at the same time they were the landlords.

Though, since independence Government of India has been taking various measures in order to improve the conditions of these down trodden people, their conditions have not yet been improved to a satisfactory level. Till today they are exploited in various aspects viz., wages, health facilities, welfare facilities, etc.

Conversely, the condition of plantation labourers may be considered better as they have permanent employment, their counterpart, i.e. the surplus plantation labourers have been put into a state of misery being unsuccessful in getting employment in plantations. Most of them are landless and live as agricultural labourers or ordinary labourers.

The health scenario of the tea labourers is not very much encouraging as they are found to suffer from various diseases, like- anaemia, Tuberculosis, diarrhoea, etc. The poor conditions of health of the tea labourers are result of diversified factors like prevalence of large-scale illiteracy. social evils viz., drinking alcohol, gambling, social culture and customs, standard of living, low dietary quality, in hygienic habits, etc. For improving the conditions of health there are various provisions in The Plantation Labour Act, 1951. In spite of these provisions their health conditions have not been improved to the desired level. Moreover, it has been found that these provisions are not properly implemented in the gardens. Though hospitals are there in the gardens but several discrepancies like- poor sanitary arrangements, lack of enough stock of prescribed medicines, non-availability of separate examination rooms, lack of sufficient beds, etc. are prevalent in the garden hospitals. Also, the number of medical and auxiliary personnel and number of beds in the Dewan hospital is not according to the scale mentioned in the Act. Besides, the number of trained nursing attendants is not as per the Act, in all the four tea gardens.

Moreover, as far as availing of the medical facilities by the garden labourers is concerned it is revealed through survey that majority of the garden labourers are still having faith on traditional health system and till today, many of the labourers believe that diseases are result of evil spirits and effect of evil stars. The garden labourers generally approach the garden hospital when the traditional treatment fails.

With regard to crude death rate, it is found to be the higher in the tea gardens under study than that of the people of Cachar. Assam and India. This signifies poor or lack of adequate medical facilities and awareness among the labourers of the gardens towards modern means of medical treatment. Besides, the labourers are mostly affected by anaemia/chronic weaknesses. The Senior Medical Officers of all the tea gardens are of the opinion that consumption of country liquor is a major cause of such type of diseases. Anaema is generally diagnosed when the haemoglobin percentage is below 70%. But among the anaemic patients in case of the tea garden labourers it is very often below 50% (Tea Garden Hospital Records). This has been to be more with the women because of repeated pregnancy. Among the men folk it is caused mainly from the habits of drinking of indigeneous liquor. Because, gastritis follows drinking which in turn prevents iron absorption, ultimately leading to anaemia.

In case of Infant Mortality Rates (IMR), IMR is highest in Pathemara Tea Garden followed by Pallorbund, Doyapore and Dewan. Less number of IMR in tea gardens shows a better picture of health facilities in the garden hospitals as well as improved socio-economic conditions of tea labourers. In comparing the IMR of the tea gardens under study with that Cachar, Assam and India, tea gardens under study are found to have the lowest, followed by Cachar, Assam and India. It has been found that the crude death rate is higher in the sample units as compared to India, Assam and Cachar district. This acts as an evidence that the health facilities in the sample tea gardens are not up to the required standards.

in case of the doctor population ratio, tea garden labourers are better than the general population of Cachar but lagging behind the general people of Assam. In case of nurse-population ratio sample tea gardens are worse off than Cachar and Assam. Again, in terms of Bed – Population ratio, the tea garden labourers of sample tea garden are better than general population. In terms of population per health centre/sub-centre we find that the tea garden labourers are in advantageous position than the people of Cachar district and Assam. In this way, as a whole it can be inferred that the health facilities of tea garden labourers are not at par when compared with Assam and Cachar as whole.

The majority of the tea garden labourers are found to be living in the Bamboo Structure or Mud houses where they could hardly accommodate themselves. As a result, they have to live in a limited space and congested environment, which ultimately affects their health. In view of limited space in the houses, most of the dwellers have made small huts used as kitchens. The labourers very often keep domestic animals either in their verandas of houses or build small sheds attached to the houses. This again adversely affects their health.

Also, with regard to hygienic habits, it has been observed that the labourers are not habitual of cleaning teeth, bathing and washing clothes daily and regularly. They clear their teeth with charcoal powder or a twig of neem tree or of a tea bush. They do not bath in using clean water or wash their clothes regularly but do according to their convenience or availability of water. Most of the labourers defecate in the open space, which is a cause of intestinal diseases in the garden. Generally, they defecate beside the road or the bush weeds thereby. The remaining labourers also do not have proper latrines and urinals. Moreover, toilets that are constructed in such a way as not to require daily cleaning. Thus, both defecating in the open space and the kachha (non-sanitary) toilets contributed more or less, in hygienic conditions in the tea gardens. In the area of the plantations also there is not even a single toilet or latrine for the use of the labourers.

Despite the availability of the medical facilities, the labourers in the sample tea gardens have been depending on the superstitions for the cure of the diseases. The labourers of the tea gardens are still afraid of visiting the Sickmaan (Hospital) and like to escape from injection and saline. Sometimes the indoor patients run away from the hospital due to the fear of *Sui* (injection) and saline.

As regards the educational status of the tea labourers, a satisfactory picture cannot be drawn, as the educational level is low. During the colonial period planters paid a very scanty attention to educate the workers and their children despite of the recommendations made by the Royal Commission on Labour in 1931. However, after the attainment of independence, National Government through the Plantation Labour Act, 1951, made it compulsory to establish garden level schools for primary education at the cost of the management. So, as per the Act the schools were established in the gardens under the control of the garden authority Although in all the tea gardens under study primary schools were established before the enactment of The Plantation Labour Act, 1951. Gradually, after independence the State Government took up these schools. But till today the parents are reluctant to send their children to the schools. Since the children are profitable both to planters and the parents alike, they are deprived of further education after primary level or sometimes before that also. As a matter of fact, the parents are unable to bear the stage wise increasing expenditure in the schools.

As regards the literacy rate of the tea labourers in the sample tea gardens, It is found to be the highest in Doyapore tea garden (63.51%). Now, if we look at sex wise, literacy rate is higher in case of males as compared to females. The wide prevalence of illiteracy among the women points to the existence of conventional attitude towards the women education. Again, the percentage male literacy is highest in Dewan Tea Garden and highest female literacy is found in Pathemara Tea Garden. The lowest percentage of male literacy is found in Pallorbund Tea Garden and also the lowest female literacy is found in Pallorbund tea garden. The overall literacy rate in sample tea garden is found to be 63.02%.

As regards the educational level of tea labourers it is found that most of the labourers are educated up to the primary or middle school level. Very less number labourers are reaching up to the H.S. level or graduation or post-graduation level. This also brings out the lack of institutions providing education up to graduation or post-graduation. There were no colleges situated adjacent to the garden area.

Regarding the use of communication media, most of the households are having a T.V. Also, a good number of labourers are found to use keypad mobiles. Some of the labourers are also found to using the android mobiles. These means of communication have wide implications with regard to exposure of the plantation labourers towards the external world. Exposure to this is how far consequential for their development is the question to be researched out.

The overall percentage of tea labourers reading newspapers are found to be around 47 percent which is higher than the reading habits of rural households of India which was 31% in 2017 (www. Statista.com). The low profile of the habits of reading newspapers is contingent upon their educational and economic levels. The number of the newspapers read by the respondents, sources of their availability and the diversifying economy of the gardens, all have their roles in the ongoing process of habit formation for newspaper reading among the tea labourers.

Digitization meaningfully affects the life of a person by rationalization processes, enhancement of efficiency, and providing prompt access to information and services. It has been evidenced that the exposure of tea labourers towards digitisation does not improve their awareness towards health and educational aspects among them. Hence, in this context proper awareness must be initiated so that the tea labourers become aware about the positive aspects of digitisation towards the improvement of their health and educational scenario.

The provisions of welfare of plantation workers relevant to health and education as mentioned in The Plantation Labour Act, 1951, it has been found that management of the tea gardens included in the sample is not properly implementing these provisions. Some provisions are strictly violated while some provisions are partially implemented. For example-Canteens (Section 11), Crèches (Section 12), Recreation Centres (Section 13), Welfare Officers (Section 18), are not there in the tea gardens under study. It means that these sections of The Plantation Labour Act, 1951, are totally violated in those tea gardens.

The facilities for protection from weather as mentioned in Rule 71, prescribed under Section 17 are also not provided to labourers in the tea gardens under study. The labourers of these four tea gardens are totally deprived of such facilities.

Section 19 with regard to working hours is also not properly implemented. The provision of weekly holidays as mentioned in Section 20 is also not properly allowed to the workers. Daily intervals for rest as mentioned in the section 21 are properly implemented in the gardens Moreover the provisions of section 22 regarding spread over are more or less implemented in the gardens.

With regard to the implementation of section 30 ie annual leave with wages are implemented in the tea gardens under study but with some discriminations as adolescent workers are very often deprived of such facilities. Sickness and maternity allowances as mentioned in section 32 of the Act are also not properly implemented in the tea gardens

Thus, we find that the welfare provisions are not properly implemented in the tea gardens under study. Also, due to the liberal trade policy by the Government of India, there has been a number adverse consequences on the tea industry, like, more competition due to arrival of cheap tea in Indian market, reduction in price and profit, more debt burden, uncertainty in tea trade, etc. As a consequence, the management of the tea gardens are not able to afford all the facilities for the tea labourers as mentioned in the Plantation Labour Act, 1951.

Also, in the tea gardens very often discrimination is made between the permanent labourers and casual labourers with regard to provision of various facilities e g., medical facilities, sick leave, etc meant for the workers as mentioned in the Act and rates of wages. This type of discrimination should be abolished in the tea gardens and the casual labourers should be considered as far as possible at par with the permanent labourers in the context of welfare facilities. Otherwise, this may endanger the work culture existing in the gardens in the near future.

Suggestions:

On the basis of foregoing discussion, following suggestions are forwarded to improve the health, sanitation and educational scenario of the tea plantation workers.

(1) Simply by enacting some legislations the health conditions of the large chunk of tea plantation workers cannot be improved but by educating them in proper perspective of the problems which they are to encounter in their day-to-day life.

(2) Young boys and girls from the labour community may be given training in Public Health and Nursing respectively and they may regularly visit the labour lines and advise the people on health issues and ways to maintain hygiene and to maintain good health. This will also lead to many other positive consequences.

(3) As one of the worst features of the labourers health condition is the consumption of country-made liquors, opium and ganja. As a matter of fact, drinking has become a common practice among the tea labourers. So, the labourers should be encouraged to dispense with such evils by providing them necessary information and knowledge on the matter. In this way, the rates of absenteeism, low productivity of the labourers can be eliminated and the quality of work life can be sustained.

(4) Besides effective arrangements should be made to run some awareness programmes on health in the tea gardens. In this context, the help of college and university students may be sought during their vacations and after their examinations. The social workers may also be induced to take up this work very earnestly.

(5) The management of the tea gardens must make it compulsory for the garden hospital staffs to visit the labour lines at regular intervals, make proper arrangements for the supply of prescribed medicines in the garden hospital, improve the quality of food stuffs provided to in-patients in the garden hospital, etc for improving the health status of the labourers as sound health of the labourers is important for reducing absenteeism and increasing productivity in the gardens.

(6) Some social workers and NGOs may arrange yogic exercise classes in the tea gardens. Now there is no doubt regarding utility and efficacy of yuga therapy which can cure many common ailments.

(7) As allopathic medicines are becoming costlier day-by-day, measures may be initiated by the concerned authorities in the gardens to popularize the use homoeopathic and ayurvedic medicines.

(8) Attempts must be initiated by the garden authority to provide and maintain proper sanitation facilities for labourers both in the area of plantation and labour lines.

(9) One of the most notable features unveiled by the study is that the labourers in the gardens are very much conservative with regard to women education. In this context measures must be undertaken in the tea gardens to bring necessary changes. Labourers must realize the importance of women education and for this very often awareness camps can be organized in the tea gardens.

(10) Attempts should also be made to provoke interest among the labourers to develop savings habit This may ultimately help the labourers to spend for the further education of their children. In this context the use of Recurring deposits (RDs) and Fixed Deposits may boost up their savings in the positive ways.

(11) Management of the tea gardens should make some effective arrangements to run night schools in the tea gardens.

(12) Awareness camps should be organised in the tea gardens to make the labourers conscious about the beneficial effects of education.

(13) Management of the tea gardens must also think of introduction of vocational training courses in the schools with due permission from the appropriate authorities so that the children in the school become ready to be self-sufficient in their future life. Otherwise, the educational scenario of the tea gardens cannot be upgraded.

(14) School curriculums should be developed in such a way that it must not become boring for the children and be related with real life. Moreover, various heterogeneous facilities like-mid-day meal, books, etc should be uniformly dispersed in all the schools of the tea gardens.

(15) The concerned authorities should take adequate steps to organise awareness programmes/training courses/ orientation courses for the teachers of the schools in the gardens so that the teachers can update their existing knowledge base and as a result the students are also benefitted.

(16) The labourers should be encouraged as not to treat their progenies simply as an instrument of earning money or economic assets rather they should send the children to the schools for further education.

(17) The provision of providing free education to the children up to 12 years of age by the management of the tea gardens may be increased to 16 years so that Middle School and High School may be included. At present the education up to eighth standard has been made free and universal by the Government.

Foregoing discussion makes it clear that though the provisions are there, but until and unless proper steps are taken for their implementation, the dilapidated condition of the labourers cannot be improved. Also, along with other problems the labourers are getting lesser wage in comparison to workers in other industries. As a consequence, though the workers are enjoying some of the facilities as per The Plantation Labour Act, 1951, they are still in distress, backward and poverty stricken, the main causes of which are large-scale illiteracy, ignorance, lack of guidance, the habit of drinking alcohol, etc. So proper steps must be initiated to overcome these drawbacks alongside the proper implementation of the provisions of The Plantation Labour Act, 1951 otherwise the future of the tea gardens will be jeopardized in the coming future.

******

Bibliography

A joint project of the United States Agency for International Development (USAID), Nathan Associates Inc., and J.E. Austin Associates (2002) "A competitiveness Strategy for Sri Lanka's Tea Industry," Developed Sri Lanka Tea Cluster, Colombo, Sri Lanka.

Allen, BC, et al., (Reprint 2001), *Gazetteer of Bengal and NorthEast India,* Mittal Publication New Delhi, India

Annual Report on Labour Immigration into Assam, Assam Secretariat Press, Shillong, 1879, p 24, 1880,p 19, 1883, p. 13, 1884, 1885, 1887, 1889, p.51. 1890, 1892, p.60, 1993, p.68, 1894, p. 70, 1895, p. 76, 1896, p. 47, 1900, p.26; 1901, p. 14, 1902-3.p. 13, 1903-4, 1904-5, Report on Immigrant Labour in the Province of Assam, Secretariat Press, Shillong, 1932,1933. Report on Tea Culture in Assam, 1889, Assam Secretariat Press, Shillong, 1890, pp.1-2; 1893. P.2.

Ariyawardana, A (2003) "Sources of Competitive Advantage and Firm performance: The Case of Sri Lankan Value Added Tea Producers", Asian Pacific Journal of Management, 20:73-90.

Arya, N., (2013), "Indian Tea Scenario", *International Journal of Scientific and Research Publications,* 3(7), (July).

Ashby, H.K. (1977), *Cocoa, Tea and Coffee,* Priory Press Limited, Hove, Crane, Russia and Company Inc., New York.

Asopa, V.N. (2004), *Competitiveness in Global Tea Trade*, Oxford and IBH Publishing Company Pvt. Ltd. New Delhi, India.

Assam Plantations Labour Rules, 1956 with The Plantation Labour Act 1957- Orient Law House, Allahabad, India, 1991.

Athparia, R.P., (2009), "Settlement and Socio-Economic Life of the Tea Garden Labourers in Cachar District of Assam", in Sengupta, S. (ed.), *The Tea Labourers of North-East India,* Mittal Publications, New Delhi, India.

Awasthi, R.C. (1975), *Economics of Tea Industry in India,* United Publishers, Guwahati, India.

Baak, Paul E. (1992) "Planter's Lobby in late 19th Century, Implications for Travancore," Economic and Political Weekly, August 15.

Balgir, R.S., (2009), "Health and Morbidity Pattern in the Tea Garden Labour Communities of North East India- A Brief Overview", in Sengupta, S. (ed.), *The Tea Labourers of North-East India,* Mittal Publications, New Delhi, India.

Banerjee G.D. and Banerjee S. (2008) "Global Tea Trade: Dimensions and Dynamics", Abhijeet Publications, New Delhi, India.

Barkatakı, S., (1969), *Assam,* National Book Trust of India, New Delhi

Baruah, P., (2009), "A Study of Heath, Hygiene and Sanitation among the Tea Garden Labourers of Assam- A Case Study of Jamira Tea Garden", in Sengupta, S. (ed.), *The Tea Labourers of North-East India,* Mittal Publications, New Delhi, India.

Bhadra, R.K. (1997) "Some Features of Tea Plantations and Surplus Labourers in Assam" in Bhadra, R.K. and Bhadra, M (edn.), Plantation Labourers of North east India, N.L. Publishers, Dibrugarh, Assam.

Bhadra, R.K.: "Health Culture of the Tribal Workers of Tea Industry-Change and continuity" in Manish Kumar (ed.), Tribal India-Problems and development Prospect (Vol. II). Gain Publishing House, New Delhi, 1996.

Bhadra, R.K.: "Some Features of Tea Plantations and Surplus Labourers in Assam" in. Bhadra, R.K and Mita Bhadra (ed.), Plantation Labourers of North east India, N.L. Publishers, Dibrugarh, Assam, 1997.

Bhattacharjee, J. B., (1977), *"Cachar under British Rule in North East India"*, Radiant Publishers, New Delhi.

Bhattacharjee, J.B (1977) "Cachar under British Rule in North East India", Radiant Publishers, New Delhi.

Bhattacharjee, Tonmoy, (2003), *Political Economy of North East India,* NICETY Graphics, Silchar, Assam,

Bhowmik, Sharit Kumar (1988), 'Ideology and the Co-operative Movement, Worker Co-operative in the Tea Industry,' *Economic and Political Weekly,* Vot. XXIII, No. 51, December, 17.

Bose, Kunal, (2015), "Working conditions in Assam Tea Grdens Poor", *Business standard,* dated: 25th September.

CCSU, (2009), Cachar Chai Shramik Union Office Records.

Chakraborty Sanjay: *Social and Cultural Change in the Ex- Tea Garden Labourers of Durgakuna in Cachar District*, Unpublished Dissertation submitted for partial fulfillment of M. Phil Degree, Department of Sociology, Assam University Silchar, 2000

Chakraborty, Dibyendu (1997) "Tea Industry: Over Supply Potential Threat," Commodity Focus, *Journal of PTI Economic Service,* District Industrial Centre, Thiruvanthapuram.

Chakraborty, M (2008) "Health and Educational Scenario of the Tea Plantation Workers in Cachar District of Assam in the Context of The Plantation Labour Act, 1951", Unpublished Dissertation, Department of Commerce, Assam University, Silchar, Assam.

Chakraborty, M (2012) "Profitability of Tea Gardens: A Study in the Barak Valley Region of Assam", in *Source- A Peer Reviewed Journal of Humanities and Social Sciences,* Vol-II, pp123-133.

Chakraborty, M. and Ahmed, J.U., (2016), *Institutional Financing of Tea Industry in India*, Lap Lambert Academic Publishing, Germany.

Chakraborty, M. and Barman, M., (2015), "Look East Policy and Tea Trade: An Evidence from the Indian Tea Industry", in Bhattacharjee, A. (ed.), *Look East Policy: Perspectives from the South-East Asian Architecture*, Excel India Publishers, New Delhi, India.

Chakraborty, R.N. (1997) "Socio Economic Development of Plantation Workers in North East India", N.L. Publishers, Dibrugarh, Assam, India.

Chakraborty, R.N., (1997), *Socio Economic Development of Plantation Workers in North East India,* N.L. Publishers, Dibrugarh, Assam.

Chattapadhay. Kanailal, (1989), *Asamer Cha-Kuli Andolan,* Pyapiras Publishers, Kolkata, India

Chaudhury, T.K., (1982), *Demographic Trends in Assam 1921-71,* B.R. Publishing Corporation, Delhi

Choudhury, R. K. (2006) "India's Tea Industry and Assam" in *Access North East (NENA),* Vol.3, Issue No.15.

Das, G.R. (2008) "Impact of Globalisation on Tea and Silk Industry of Assam", in Deb, B.J., Sengupta, K. and Datta Roy, B. (ed.), *Globalisation and North East India,* Concept Publishing Company, pp. 294-304.

Das, P.P (2006) "Borake Chashilper Mrityuyatrar Itikotha" (Bengali) in *Samoik Prosongo, Dated - 5th November ,* No. 200.

Das, Partha Pratim: "Borake Chashilper Mrityuyatrar Itikotha" in Samoik Prasongo. Dated-5th November 2006, No. 200

Dasgupta, B. (2008) "Impact of New Economic Policy on Tea Industry of Assam" in Deb, B.J., Sengupta, K. and Datta Roy, B. (ed.), *Globalisation and North East India,* Concept Publishing Company, pp. 305-312.

Deb, Uttam, Roy, Ashish Taru and Mazumder, Ritwik (2012) "Total Factor Productivity Growth in Selected Tea Gardens of Assam," *Source, A Peer Reviewed Journal of Humanities And Social Sciences,* Vol.II, August, pp.151-165.

Debnath Ruma: *Social and Cultural Life of the Labourers in Dewan Tea Garden,* Unpublished Dissertation submitted for partial fulfillment of M Phil Degree, Department of Sociology. Assam University Silchar, 2000.

Debnath, Ruma (2000) "Social and Cultural Life of the Labourers in Dewan Tea Garden", Unpublished Dissertation submitted for partial fulfillment of M. Phil Degree, Department of Sociology, Assam University Silchar.

Devi, K., (2022), "Socio-Economic Condition of Assam's Tea Garden Workers", *World Affairs: The Journal of International Issues,* 26(2), pp. 60-75. ()

Devi, P., (2014), "Socio-economic Status of the tea Garden Women Workers in Assam: A case Study with Special Reference to Sonitpur District", *Reviews of Literature,* Vol. 2, Issue 2, September.

Dutta, D.: "Beginning of Tea Industry in Cachar"; Purbasri, January 1965.

Dutta, M.L (1983) "Education and Employment of Tea Garden Labourers of Assam with special reference to the District of Sibsagar", unpublished Ph. D. Thesis, Gauhati University. Guwahati, Assam.

Dutta, M.L. *Education and Employment of Tea Garden Labourers of Assam with special reference to the District of Sibsagar*, unpublished Ph D. Thesis, Gauhati University. Guwahati, Assam, 1983.

Dwibedi, H.N. (1999) "Production of Tea in India, with special reference to Dooars," K.P. Bagchi and Company, Kolkata.

Feder. E "Latifundia and agricultural labour in Latin America" in T Shanin (ed), Peasants and Peasants Societies, Penguin Books, 1971, p.83

Ghosal, Sutanuka (2009) "Ageing Tea Bushes is a Major Concern" in *Economic Times,* Dated: 17th September.

Goradia, Praful (1979), *Profiles of Tea,* Oxford and IBH Publishing Company, New Delhi.

Goswami, Atul (2002*)* "Tea Industry in Assam: Its linkages with states Economy", in *Social Change and Development,* October, P-51.

Goswami, Atul: "Tea Industry in Assam: Its linkages with states Economy", in Social Change and Development (P.51), Oct. 2002

Goswami, B.K. and Dutta, P.K. (2012) "A Modified Approach to Bringing up of Young Tea into Bearing," *Assam Review & Tea New*, Vol.101, No. 8, November, pp.10-16.

Goswami, D. (1992) "A Study of the Tea Labour Population of the Brahmaputra Valley of Assam", Ph.D. Thesis, Gauhati University, pp.196-226,

Goswami, D.: *A Study of the Tea Labour Population of the Brahmaputra Valley of Assam,* Chapter V. Ph.D. Thesis, Gauhati University, pp. 196-226, 1992.

Goswami, P.J., (1995), *Tribals in Transition: A Socio-Economic Study,* Jawahar Publishers, New Delhi

Goswami, P.J.: "Problems and Prospects of Tea Industry in Assam", in A. Banerjee (ed.), Economic Reform and Vision in North East India, Kanishka Publisher, New Delhi, 2007.

Guha Thakurta, J. "Role of Women in India's Tea Industry", in Tea Association of India's News Letter, January, 1986, p. 10.

Guha, A,, (1977), *Planter- Raj to Swaraj Fredom Struggle and Electoral Politics in Assam 1826-1947,* ICHR, New Delhi. 1977, p.14.

Guha, A.: "A Big Push without a Take-off: A Case Study of Assam, 1871-1901. Reply to Comment" in Indian Economic and Social History Review, Vol. XI. No. 4, December, 1974 (a), p.475.

Harijan, K. (2011) "Role of Tea Plantation Industry in Human Resource Development in Darjeeling Hills" in *Darjeeling and the Indian East, Salesian Journal of Humanity and Social Sciences, Vol. II, No. I,* May, pp. 85-96.

Harlalka, S.S. (1975) "The Tea garden Labourers of Assam" People's Publishing House Pvt. Ltd., New Delhi.

Harlalka, S.S., (1975), *The Tea garden Labourers of Assam,* People's Publishing House Pvt. Ltd. New Delhi.

Haviland, C. (2011) "Lanka's Tea Workers Fear for Their Future" in *Seven Sisters Post*, Dated: 24th April, 2012, P. 12.

Hazarika, D., Ozah, D., Satyakam, S., Pegu, H.K. and Gogoi, P. (2024), "Socio-Economic Conditions of the Tea Garden Labourers of Dibrugarh, Tinsukia, Dima-Hasao and Goalpara Districts of Assam and Determinants of Wage", *Library Progress International,* 44(4)inhttps://bpasjournals.com/library-science/index.php/journal/ article/ view/3828 (visited: 20-04-2025).

Hazarika, K (2011) "Changing Market Scenario for Indian Tea", *International Journal of Trade, Economics and Finance*, Vol. 2, No. 4, August, pp.285-287.

Huque, M.R.H (2007) "Strategic Cost Management of Tea Indusry: Adoption of Japanese Tea Model in developing Country Based On Value Chain Analysis", in mome.lib.ynu.ac.jp/ dspace/bitstream/10131/3125/1/4-Huque.pdf. (visited 20th Dec, 2009).

Hussain, M.M. and Hazarika, S.D. (2010) "Assam Tea Industry and its Crisis," *Dialogue,* July-September, Vol.12, No.1.

ITA, (2015), Indian Tea Association, Press Release, dated: 2nd March, 2015, from http://www.indiatea.org/downloads/PR020315.pdf , (visited: 20-07-2016).

Jain, Ajith (1995), 'Drink Tea To Stay Healthy and Cancer - Free, '*The Planter's Chronicle,* Vol.90, No: 7, July.

Kar Deb, Ajanta and Mathew, Binu (2012) "Tea Industry in West Garo Hills District of Meghalaya- A Micro analysis," *Source, A Peer Reviewed Journal of Humanities and Social Sciences,* Vol.II, August, pp.93-104.

Kar, A.K. (1996): "Social Transformation Among the Migrant Tea Labourers in Assam with special reference to Tribal Workers", in Deepak Kumar Behera (ed), Contemporary Society Tribal Studies Development Issue and Change, Vol II. Concept Publication, New Delhi

Kar, R.K. (1997) "A Note on Health and Sanitation among the Tea Labour in Assam", in Bhadra, R.K. and Bhadra, Mita (ed.), Plantation Labourers of North east India, N.L. Publishers, Dibrugarh, Assam.

Kar, R.K.: "A Note on Health and Sanitation among the Tea Labour in Assam", in Bhadra, R.K. and Mita Bhadra (ed.). Plantation Labourers of North east India, N.L. Publishers, Dibrugarh, Assam, 1997.

Kar, R.K.: "A study of some aspects of Marriage in a Tea Garden Tribe of Mancotta Tea Plantation, Dibrugarh, Assam", in S.M. Dubey (ed), NE India A Sociological Study, Concept publishing Company, New Delhi-1978

Kar, RK: *Tea Plantation and its Labourers in Assam An Overview,* Paper presented in Regional seminar on Society and Development in Tea Gardens of North East with Special Reference to Barak Valley, Assam University, Silchar, 1998.

Kumar, B and Goswami, P.J.: "Marketing strategies of Tea Manufacturers in Assam: Critical Analysis" in DGCC's Journal of Commerce. Vol-iii, No 1 2006.

Kumar, Purnendu (2006) "State and Society in North East India- A Study of Immigrant Tea Plantation Labourers"; Regency Publications, New Delhi, pp. 86-93.

Kumar, Purnendu, (2006), *State and Society in North East India- A Study of Immigrant Tea Plantation Labourers.* Regency Publications, New Delhi, pp. 86-93

Kurian, Tessy (1990) "Socio-Economic Background and Consumption Pattern of Women Workers in the Tea Industry in Munnar, ldukki District," M. Phil. Dissertation, Department of Applied Economics, Cochin University of Science and Technology, Cochin.

Laskar, B. K (2001) “Tea Industry: A Need to Shift in Paradigm”, in *The Assam Review and Tea News,* January.

Laskar, B.K.: Tea Industry: "A Need to Shift in Paradigm", in The Assam Review and Tea News, January 2001.

McLeod Russel India Ltd. & Harrison Malayalam (2007) “Tea Industry”, Arm Research, (*www.armresearch.in.)* visited: 29-04-2012.

Mitra, Debabrata, (2010), *Globalisation and Industrial Relations in Tea Plantations,* Abhijeet Publications, New Delhi, India.

Mukherjee, S.R., (1967), *Plantation Labour in India,* Prabasi Press, Kolkata.

Muraleedharan, N. (1998) “A review of recent attempts on mechanisation of field operations in Tea,” Bulletin of UPASI Tea Scientific Department, No. 51, March.

Nagoor, B.H (2009) “Performance of India’s Tea Exports: A Comparative Study of Major Tea Countries of the World,” Quantitative Approaches to Public Policy, pp. 062-21.

Park, K., (Reprint 2007), *Park's Textbook of Preventive and Social Medicine,* M/s Banarasidas Bhanot Publishers, Jabalpur, India.

Raman, Ravi, K. (1986) “Plantation Labour: Revisit Required,” Official Paper, *Economic and Political Weekly,* Vol. XXI, No. 22, May 31.

Rao. C. Niranjan "Geographical Indications in Indian Context-A case study of Darjeeling Tea", in Economic and Political Weekly, Vol XL. No 42. October 15-21, 2005

Roy. Sanjib "A Review Role of Tea Industry in India", in The Assam Review and Tea News, March, 1999

Saikia, S. K. (2006) “Prospects Tea Tourism in Tinsukia and Dibrugarh Districts” in *DGCC’s Journal of Commerce,* III(1),in https://www.jstor.org/stable/48681981(visited: 20-04-2025).

Saikia, Sashi Kanta "Prospects Tea Tourism in Tinsukia and Dibrugarh Districts" in DGCC's Journal of Commerce. Vol-im. No. 1, 2006.

Sarkar, Kanchan, and Bhowmik, Sharit, K. (1988) “Trade Unions and women workers in Tea Plantations,” *Economic and Political Weekly.* December, 26.

Sengupta, S. (2009) “Tea to get Smart Makeover to Capture Young Market” in *Times of India*, Dated: 9th November,

Sharma, G., (2013), “A Case Study on Socio-Economic Condition of Tea Garden Labourers –Lohpohia Tea Estate of Jorhat District, Assam”, *The Echo,* A journal of Humanities & Social Sciences, Vol. I, Issue III, January.

Sharma, K., (2003), *Tea Plantation Workers in a Himalayan Region,* Mittal Publications, New Delhi, India.

Sharma, K., (2005), *Sociology of Indian Tea Industry,* Mittal Publications, New Delhi, India.

Sharma, Poonam, (1989), *Family Welfare Programme in India*, Deep and Deep Publication, New Delhi,

Shil, Parag (2002) “Export Marketing Practices of Indian Tea Industry: 1991-2000”, Unpublished Dissertation submitted for partial fulfillment of M. Phil Degree, Department of Commerce, Assam University Silchar.

Shil, Parag: Export Marketing Practices of Indian Tea Industry: 1991–2000, Unpublished Dissertation submitted in Partial fulfilment of M. Phil. Degree, Department of Commerce, Assam University, Silchar, 2002.

Singh, Bhupinder and Mohanty, Niti, (1995), *Tribal Health in India*, Inter Publication, New Delhi

Singh, S.N., Narain, A. and Kumar, P., (2006), *Socio-Economic and Political Problems of Tea Garden Workers,* Mittal Publications, New Delhi, India.

Subhramaniam, B (1995) “Tea in India”; Publication and Information Directorate, Dr. S.K. Marg, New Delhi and Wiley Eastern Limited, 4835/24, Dariyaganj, New Delhi.

Subhramaniam, Bala, (1995), *Tea in India, Publication & Information*, Directorate, Dr. S.K. Marg, New Delhi and Wiley Eastern Limited, 4835/24 Daryaganj, New Delhi

Sundar, P.S. (2012) "Initiatives to Encourage Quality Tea," *Assam Review and Tea News,* October, Vol.101, No 8, pp. 24-26.

Sundar, PS: "Imprudent Policy A Key Limiting Factor", in Assam Review and Tea News, Vol-96, No. 5, 26th July, 2007,

Thakurta, G.J. (1986) "Role of Women in India's Tea Industry", in *Tea Association of India's News Letter,* January, p-10.

Thapa, Ajit (2005) "Concept Paper on Study of Nepalese Tea Industry Vision-2020" Nepal Tree Crop Global Development Alliance (NTCGDA) Winrock International, Baneshwor,Kathmandu.

Varandani, G., (1989), *Social Security for Industrial Workers in India,* Deep and Deep Publication, New Delhi, India.

Verner, G., (1858), *"Letter to R. Houtson, dated: Sept 4",* in Cachar District Records, 1969, pp. 10-20

Wal Van Der, Sanne (2008) "Sustainability Issues in the Tea Sector: A Comparative Analysis Six Leading Producing Countries", in http:// www.somo.nl/html/ paginas/ pdf/Sustainability Issues_in_Tea_Sector _EN.pdf> visited: 02-05-12.

Wickramasinghe, A. D. and Cameron D. C. (2003) "Economies of Scale Paradox in the Sri Lankan Tea Industry: A Socio-Cultural Interpretation", 16th International Farm Management Congress, Perth, W. Australia, 10th - 15th August 2003, The University of Queensland, Brisbane, Australia.

Government and Other Institutional Publications:

Basic Statistics, NER, 2006, pp. 280-290.

Statistical Hand Book, Assam, 2006, Directorate of Economics and Statistics, Government of Assam.

Tea Statistics. Tea Board of India, Kolkata, 2000-2001 and 2004-05.

Census Report, 2001.

Souvenir:

Centenary Souvenir, Indian Tea Association, Surma Valley Branch, Silchar, 2001.

Acts:

The Assam Plantation Labour Rules 1956 and The Plantation Labour Act, 1951.

Websites Visited:

http://www.vedanti.com/News/assam_tea_labour_problem.htm

https://sites.google.com/site/swabhimanngo/

http://www.internationalseminar.org/XV_AIS/TS%205B/8.%20Mrs.%20Sangeeta%20Saikia.pdf

https://www.ilo.org/sites/default/files/wcmsp5/groups/public/@ed_protect/@protrav/@safework/documents/genericdocument/wcms_681831.pdf

****